Shadows of Existence

DEDICATION

For Deb, Corey, and Lauryn

…and for Ted Parker, Fonchii Chang Matzunaga, Jordi Magraner, and the other heroes of discovery and conservation

Shadows of Existence

DISCOVERIES AND SPECULATIONS IN ZOOLOGY

Matthew A. Bille

ISBN 0-88839-612-0
EAN 9780888396129

Cataloging in Publication Data

Bille, Matthew A.
Shadows of existence : discoveries and speculations in zoology / Matt Bille.

Includes bibliographical references and index.
ISBN 0-88839-612-0

1. Cryptozoology. 2. Rare animals. I. Title.

QL88.3.B54 2005 590 C2005-90492

Printed in Indonesia — TK Printing

Production: Mia Hancock
Cover design: Rick Groenheyde
Illustrations: Bill Rebsamen
Front cover illustration: Vu quang ox with Africa's okapi

Published simultaneously in Canada and the United States by

HANCOCK HOUSE PUBLISHERS LTD.
19313 Zero Avenue, Surrey, B.C. Canada V3S 9R9
(604) 538-1114 Fax (604) 538-2262

HANCOCK HOUSE PUBLISHERS
1431 Harrison Avenue, Blaine, WA U.S.A 98230-5005
(604) 538-1114 Fax (604) 538-2262

***Website:* www.hancockhouse.com**
***Email:* sales@hancockhouse.com**

CONTENTS

ACKNOWLEDGEMENTS

The writing of any book is a collaborative effort. That's especially true when the book concerns a scientific discipline involving a variety of specialists. Many people contributed in many different ways to the production of this volume.

Countless people shared scientific expertise, source material, and other information with me. In no particular order, they include Darren Naish, Dr. Karl Shuker, Dr. Cheri Jones, Peter Hocking, Dr. Robin Baird, Dr. Robert Pitman, Dr. Marc van Roosmalen, Dr. Louise Emmons, Dr. John Heyning, Dr. Janet Voight, Dr. Karin Forney, the late Dr. Grover Krantz, Mark Bayless, Dick Raynor, Timothy Isles, Lisa Bowen, John Kirk, Ben Roesch, Loren Coleman, Jane Bille, Don Bille, Chris Bille, Tim LaPorte, Richard Ellis, Larry Kanuit, Scott Norman, Bobbie Short, George Kantner, Peter Hynes, Dr. George Brown, Mark Bowen, Dr. Frank Poirier, Dr. Merel Dalebout, Dr. Charles Paxton, John Lundberg, Dr. Robert Timm, Glen Alsworth, John Moore, Michel Raynal, Angel Morant Fores, Dr. Marcella Kelly, Bruce Champagne, Jim Heffelfinger, June O'Neill, Judy Bartlett, Beth Johnson, Chad Arment, Lorelei Elkins, Chris Orrick, Craig Heinselman, Lauri Facsina, David Lupton, Peter Zahler, Ray Nelke, Henry Bauer, Dagmar Fertl, Arlene Gaal, William Gibbons, Marcelo Volcato, and Bill Kingsley. Thanks are also owed the staff of the Pikes Peak Library District, especially former librarian Kirsty Smith, for interlibrary loans and other assistance. A special helping of gratitude is due fellow authors Richard Ellis, John Kirk, Karl Shuker, Charles Humphreys, Marc Miller, William Corliss, and Loren Coleman and Patrick Huyghe for copies of their books. David Hancock provided additional books his publishing firm, Hancock House, had printed. In addition, Karl proofread and corrected two essays on primates. Richard and Darren did the same for early drafts of the whale essays, as did Chris Orrick and Warner Lew for the section on Lake Iliamna and Bruce Champagne and June O'Neill for some "sea serpent" items. I thank all these experts, along with my friend, biologist Cherie McCollough, who gave the manuscript a scientific review early in the process and ensured the accuracy of the taxonomic discussions. The greatest appreciation is owed to Bill Rebsamen, who enthusiastically tackled the job of illustrating animals based on information that sometimes consisted solely of eyewitness reports. His magnificent portraits speak for themselves.

I owe a continual debt to all those who have believed in me, encouraged me, and put up with me. That includes, of course, my wife Deb and our two little angels, Corey and Lauryn. To this list I must add friends like Robyn Kane, a constant supporter who gave the manuscript an early proofreading, and Erika Lishock, who first suggested I write this particular book. I appreciate the indulgence of my employers and colleagues at Booz Allen Hamilton and at my former company, Analytic Services Inc. (ANSER). I also owe my parents, Deb's parents, and everyone who bought my first book.

Finally, I must acknowledge the debt I owe to the Lord, who gives all of us our talents and has gifted our entire species with the curiosity that drives us to explore, catalogue, and understand our universe.

INTRODUCTION

The average layperson thinks that we know about ninety-nine percent of the species that occur on the Earth, and that's the farthest thing from the truth. On a global scale, we know maybe ten percent of what's out there.

— Gary Waggoner, Center for Biological Informatics,
United States Geological Survey.

Life has existed on Earth for over three billion years. This is a span human beings can comprehend intellectually, but are simply unequipped to visualize. For millions of generations, animal life has adapted, radiated, and evolved into every imaginable niche, from abyssal depths and boiling hydrothermal vents to the densest forests to the highest mountaintops. Animals have pushed into the hottest deserts, the coldest glaciers, and the skies above the planet. Scientists have only begun the work of cataloging the resulting web of millions of species and understanding their genetic and ecological relationships.

This book's predecessor, *Rumors of Existence*, came out in 1995. *Rumors* recounted the most important discoveries in zoology from 1930 to the date of its writing, along with rediscoveries of animals presumed extinct and a survey of reported but unconfirmed animals.

Shadows of Existence continues this scientific detective story. The essays in this book demonstrate that, in the 21st century, the search for new animals remains a vibrant, successful, and important field of endeavor. Even as conservationists race to save species and habitats from destruction, nature continues to unveil surprises. New species are emerging into the light of scientific knowledge at a rate which, in many areas, is actually increasing. Other animals—not many, but a heartening few—are being rescued from the darkness of presumed extinction, while researchers catch tantalizing hints and glimpses of creatures still in the shadows.

Most of the new animals in this book are species described from

1990 to the present. I use "described" here in the technical sense, meaning a scientist has obtained a type specimen (or holotype) of the animal and formally published a description and scientific name. Many years can elapse between the discovery of a specimen and the formal description, so some species not yet described are mentioned as well.

Concerning the animals whose existence has yet to be confirmed, I included such creatures in *Rumors* only when the case was backed up by physical evidence or reported by witnesses with some scientific training. This book broadens those limits somewhat. In the interests of sparking further inquiry and informed discussion, *Shadows* includes what I hope are fair-minded examinations of subjects which meet neither of these tests, provided the animal involved is not completely implausible on grounds of anatomy, ecology, etc., and there are enough credible reports to "keep the file open."

The pursuit of new or presumed-extinct animals includes the controversial subspecialty known as cryptozoology. This has been a problematical term ever since the late Dr. Bernard Heuvelmans popularized it over forty years ago. To Heuvelmans (1917-2001), the "father of cryptozoology," the word meant, "the scientific study of hidden animals. i.e., of still unknown animal forms…"

Why should this be controversial? First, many zoologists believe there is no need for a specific discipline devoted to new animals. The mammologists can find and describe the mammals, the entomologists can do the same for insects, and so on. Second, there is a common perception that "cryptozoology" refers only to the search (mainly conducted by amateur enthusiasts) for "monsters" like the yeti. That makes cryptozoology highly suspect, at best, as a branch of science. The cascade of books, websites, and organizations created by those cryptozoological enthusiasts who have more enthusiasm than zoology tends to reinforce this viewpoint.

While cryptozoology has been plagued by sensationalism and unsubstantiated claims, the outright rejection of this subject is nevertheless unwarranted. There are two reasons.

First, the work still to be done in finding new species is enormous, and help from any quarter should be appreciated. There is a tendency among the public and even some experts to assume that the animals, or at least the "interesting" animals—mammals, birds, sharks, etc.—have been almost entirely classified. As this book demonstrates, that is most definitely not so.

Second, there is a danger that some subjects may be dismissed without a scientific examination. For example, very few primatologists spend much time thinking about the possibility of an unclassified ape haunting the high mountain forests of the Himalayas. It's true that the yeti's existence remains unproven and, indeed, appears increasingly unlikely as decades pass without meaningful new evidence. However, it's also true that a subject like this may not be properly investigated because of the surrounding aura of "pseudoscience."

One other reason commonly heard for rejecting cryptozoology is that the large animals, at least, should have left fossil records. Perhaps so, but we have not found all the extinct animals any more than we have found all the living ones. We had no evidence for one of most startling anthropological discoveries of all time, the three-foot-tall Flores people, until the very delicate unfossilized bones, 18,000 years old, were dug out in 2004. We had—and still have—no paleontological evidence for Southeast Asia's 200-pound saola, discovered as a living species in 1992. For that matter, we have no such evidence for chimpanzees. Fossils can prove a species once existed, but their absence does not prove the contrary.

Getting back to the question of cryptozoology vs. zoology, I suggest here that the distinction often drawn between them is invalid. The methods of searching for new species—library and museum research, discussions with indigenous people, and fieldwork—are the same no matter what kind of animals are being looked for or who is doing the looking. The work, not the terminology, is what matters.

As we become ever more aware of the interconnectedness and fragility of life on this planet, it becomes all the more important to know about the animals we share it with. Conservation cannot succeed unless we know what animals live where, what their needs are, and how they affect their ecosystems. Accordingly, it is vital that we continually strive to identify new species, ascertain the status of the species we know about, and act quickly when we discover animals which need protection.

To help readers evaluate the "unsolved" cases for themselves, complete references for the information presented are listed at the end of the book, along with a collection of recent books, articles, and Internet sites on zoology and cryptozoology. Also appended is a note on taxonomy for readers who want to review the definitions of species, genus, and related terms.

So settle back, open your minds, and let us journey into the farthest realms of zoology.

— Matt Bille, November 1, 2005

SECTION I: NEW CREATURES

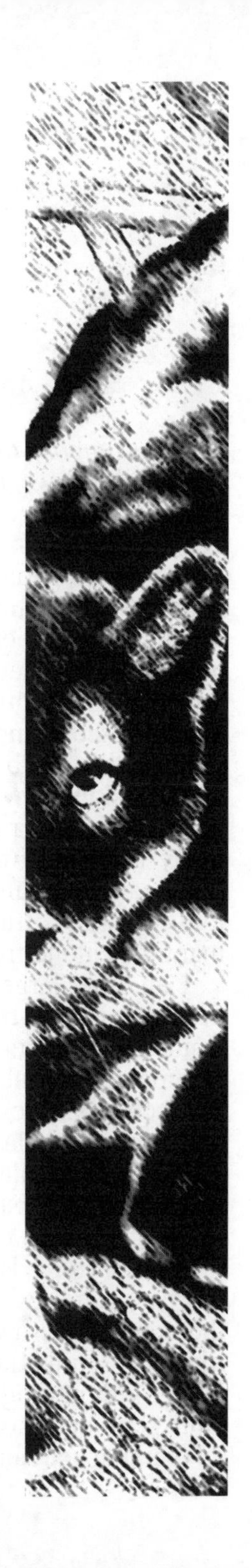

INTRODUCTION: SECTION I

Can we suppose that we have at all exhausted the great museum of nature? Have we, in fact, penetrated yet beyond its ante-chambers?

— Charles Gould, *1884*

When Gould wrote those words, he knew that a great deal about the natural world remained to be discovered. It would doubtless surprise that prominent naturalist to learn his questions are still pertinent 120 years later. Many people, even scientists, fail to realize how many new animals are still being discovered all over the globe. While insects and other small invertebrates dominate the new species, they are by no means the only creatures joining the zoological lists. Just in the 1990s, descriptions were published of two beaked whales, a jellyfish a yard across, several monitor lizards, many tropical birds, five species of large hoofed mammals from Southeast Asia, and a bewildering collection of monkeys from the rain forests of Brazil. This came on top of the 160 species of new mammals alone from the 1980s.

Can other large animals still lurk unconfirmed at the borders of our knowledge? It's hard to say "no" if one considers an example from 2002, when some very odd apes were reported in the Democratic Republic of Congo. These apes looked much like oversized chimpanzees and had a chimplike diet heavy in fruit, but they had some gorilla-like facial characteristics and nested on the ground like gorillas.

Conservationist Karl Ammann found these robust primates by investigating an oft-overlooked report made by a Belgian army officer in 1908. Hunters in the village of Bili told Ammann of a large, ground-dwelling chimpanzee they considered distinct from the normal tree-dwelling type. Ammann found nests used by these mystery animals and collected skulls, footprints, hair, and feces, along with photographs of one living specimen and one dead one. Both skull and footprint measurements were well out of the accepted chimpanzee range. Primatologist Shelly Williams, who has seen the Bili apes in the wild, commented, "At the very least, we have either a new culture of chimps that are unusually large or hybrids with unusual behaviors."

DNA sampling so far indicates the animals are indeed chimpanzees, but they represent a very distinctive population, and thus are of great scientific interest. New species or not, it is a striking and very important fact that a group of large apes could remain essentially unknown until the 21st century.

To offer a few examples from the smaller end of the scale, two beetles netted recently were so strange that, in 2002, entomologists had to create the first new beetle family erected in 150 years. Stranger still, the two species in the family Aspidytidae don't live anywhere near each other. One is from South America and the other from China. (One of the venerable truisms of science is that some discoveries create more questions than they answer.) Also in 2002, the first new order of insects since 1914 had to be created to house Africa's bizarre gladiators, which look like heavily armed grasshoppers about an inch long.

"Small," of course, is a relative term. A paper published in 2001 described a new spider from specimens nabbed in a Laotian cave in 1933 and overlooked in a Paris museum collection for several decades. How it was overlooked is a bit of a puzzle. *Heteropoda maxima* is one amazing spider, with a legspan of up to twelve inches.

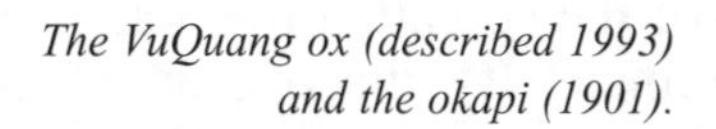

The VuQuang ox (described 1993) and the okapi (1901).

Then there are the oceans. Experts have estimated that one-third or more of the creatures of the seas have yet to be described. Most new finds will no doubt be tiny invertebrates, but the waters of the Earth are vast, and some large animals are still out there.

We know from photographs and expert observations there is at least one species of whale awaiting classification. Also awaiting a definitive name is the bizarre squid, over twenty feet long and with huge terminal fins and long, slender tentacles, recently spotted by submersibles and robotic exploration vehicles at depths of 6,000 feet or more in three widely separated spots in different oceans. Sometimes we even find a new species already in hand: a bizarre shark two feet long, which has red bristles on its skin, oversize teeth, and uses its fins in a "hopping" motion, turned up in a shipment to a German aquarium in 2004. It apparently had spent a few years in captivity with a series of owners, and scientists are trying to trace its history and figure out where it was collected. The specimen has not yet been scientifically described. For now, according to media accounts, it has been saddled with the peculiar nickname of "Cuddles."

The idea that we know all the significant inhabitants of the natural world is, to put it bluntly, absurd. There is every reason to keep up the search for new species, both in remote areas of the globe and in places we supposedly know but might not have explored completely. If the last decade is any guide, we are guaranteed to find some truly unexpected and amazing species, large and small, as we continue the enormous task of cataloging the animal kingdom.

NEW GUINEA'S NEW KANGAROO

In 1994, a striking new mammal, one of the so-called tree kangaroos, was confirmed from New Guinea. Its discoverer was Australia's best-known mammologist and paleontologist, Tim Flannery.

The balding, bearded Flannery is Principal Research Scientist at the Australian Museum in Sydney. He was already known for coauthoring the 1990 description of Scott's tree kangaroo (*Dendrolagus scottae*), which he helped discover. Scott's tree kangaroo is New Guinea's largest known mammal.

Flannery described the newest find, locally called the dingiso (pronounced din-gee-soh), as "very primitive in its body plan and behavior." His point was that the animal is not particularly specialized for tree life. In fact, it spends most of its time on the ground, where it has less competition from other species. As Flannery put it, "The dingiso is a tree kangaroo that doesn't live in trees. The rea-

The reported black and white jaguar of Brazil.

An unusual reverse-striped tiger from India.

The mystery cat, called mngwa *or* nunda, *feared in East Africa.*

Sasquatch — a North American great ape?

Steller's sea cow, believed hunted to extinction in the 18th Century.

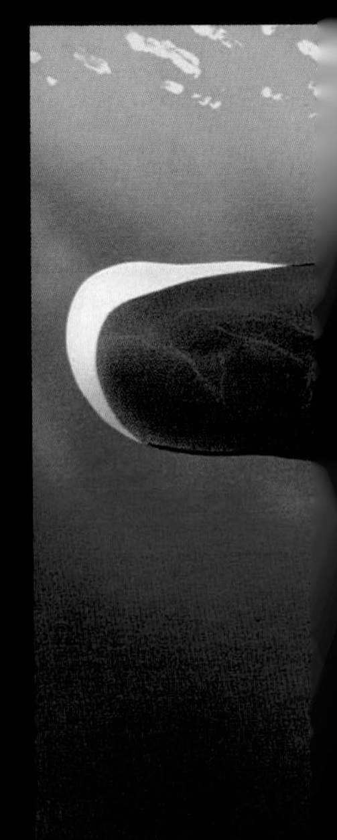

Many strange fish were reported by pioneer ocean explorer William Beebe.

Aerial sighting of the mysterious fish-like giants of Lake Iliamna in Alaska.

The strange-looking manta reported by William Beebe and others.

The mysterious beaked whale, Mesoplodon "Species A," with other beaked whales.

The mimic octopus and its relative, the "Wonderpus."

Even in the 21st century, sailors and divers occasionally report strange "sea serpents."

Africa's alleged surviving dinosaur, the mokele-mbembe.

The Vu Quang Ox, shown here with a famous discovery from 90 years before—Africa's equally strange okapi.

son it doesn't live in trees is most tree kangaroos live in forest. The dingiso lives right up under the glaciers."

Those glaciers are in Irian Jaya, the Indonesian province that includes the western half of the world's second-largest island. It's a good place for animals to hide. Irian Jaya encompasses over 160,000 square miles. The indigenous people, the Papuans, are split into hundreds of tribes, each with its own customs and language. That diversity, along with the mountainous, heavily forested terrain of the interior, makes thorough exploration a devilishly difficult business.

Flannery described the environment to an interviewer by saying, "...you look at those mountains that just go up and up and then there's another great series beyond that right up to the grasslands above the tree line. And those places just haven't been disturbed. There's no extinctions in them yet. And there's wonderful undiscovered fauna."

So it was that, until recently, science was ignorant of the dingiso (now formally known as *Dendrolagus mbaiso*). The new kangaroo can weigh over twenty pounds and is black with a white chest and other white markings, including a blaze on its face.

Flannery's search for the creature began in 1990, when he obtained a Papuan headdress made from the skin of an unidentified kangaroo. That same year, Gerald Cubbitt photographed a captive juvenile of the species. It took three more years to track down and examine adult specimens. The scientific paper naming the species was published in 1995.

The dingiso shows no fear of people, emits a whistling call, and is altogether adorable by human standards. That last trait may be significant. The cause of conservation can always use a new "poster animal." Saving habitat for the tree kangaroo, as Flannery hopes will happen, will save it for countless other species—known and unknown.

FISHING IN SOUTH AMERICA

The discovery of a new species of fish is not, generally speaking, particularly newsworthy. The flood of new species from three recent South American expeditions is another matter. These endeavors illustrate how much there is to learn even about well-traveled bodies of fresh water.

In 1996, a team led by Dr. John Lundberg of the University of Arizona, funded by the National Science Foundation, set out to trawl the deeper channels of the Amazon River. While much of the river is heavily fished, Lundberg suspected that a great deal of aquatic life was active well below the surface, safe from the nets and lines of fishermen. Lundberg's team dragged its equipment through water thirty to a hundred feet deep and hauled up 240 species, many never seen before.

Examination of the catch showed that most new species in these muddy, nearly opaque waters are either catfish or electric fish. Some of the new species discovered would satisfy anyone's taste for the bizarre. There are two species of electric fish which eat nothing except the tails of other electric fish. There is an electric fish with a tonguelike projection of unknown function above its chin, and there are two electric fish which appear identical but are separate species—distinguishable only by their differing patterns of electrical discharge. There are transparent catfish, armor-plated catfish, and one incredibly tiny species, one-third of an inch long, which has both these characteristics. (Catfish species seem to be proliferating in icthyological records like algae in an unkempt aquarium. Counting the diminutive Amazon species, there are over 2,800 known catfish species, and American ichthyologist Larry Page estimates 1,000 to 2,000 remain to be classified. When the famed biologist Archie Carr wrote in 1941 that "any damned fool knows a catfish," he had no idea how many there were to know about.)

Even after Lundberg's effort, not all of the Amazon system has been explored. One stretch of the Rio Negro is over 300 feet deep, beyond the reach of the Lundberg expedition's nets. There is no question it holds more new species. But what species? And who will find them?

The Lundberg expedition's news was not the only recent icthyological bulletin from this continent. Also in 1996, the U.S.-based Conservation International (CI) sent its first AquaRAP (Aquatic Rapid Assessment Program) team to the Tahuamanu and Manuripi River basins in Bolivia. Over twenty scientists spent a month netting specimens in rivers, streams, and swamps. They produced five new species of fish, the most spectacular described as "a dinosaur-like armored catfish covered in bony plates."

CI scientists found ten more new fish in Venezuela's Upper Caura River late in 2000. The first formal description published as a result, in March 2003, named a two-inch-long green and silver fish with a striking red tail the bloodfin tetra (*Aphyocharax yekwanae*).

Overall, about 200 new fish species are added every year to the 25,000 already known. In January 2000, a new species of bass, *Micropterus cataractae*, was even described from well-known and commonly fished rivers of the southeastern United States.

Fishermen had known of the "shoal bass" for at least fifty years, but no scientific examination had been made to see if this fish differed from other bass. A large shoal bass can be two feet long and weigh over eight pounds. There are a lot of surprises left in the ponds, lakes, and rivers of the world.

RAP ADVENTURES

In the last essay, I mentioned the work of an AquaRAP team from Conservation International (CI). Since 1990, RAP and AquaRAP teams have been deployed in biological "hot spots" around the world. The objective of these visits is to spend a matter of weeks making an intensive survey of the flora and fauna in areas subject to imminent threats, such as development. The work of RAP teams has contributed to establishment of six new protected areas in five countries.

The scientific results have been impressive. In addition to documenting the biodiversity of regions we knew little or nothing about, CI researchers have found new species everywhere they went.

Three more examples of RAP expeditions will illustrate the scope and productivity of this work. In July 1999, CI announced the results of a 1996 survey of Papua New Guinea's Lakekamu Basin. The RAP team netted twenty-two new species of ants, bees, and wasps along with eleven new frogs, seven new lizards, and three new fish. Ornithologist Andrew Mack, who led the expedition, commented that, "It's clear from the large number of new species discovered during just one month of survey that there's an urgent need for more biological inventories and taxonomic studies in this region." Ominously, he added, "While we continue to identify the species collected in the Basin, forests surveyed by the RAP team are already being logged."

An AquaRAP team visiting Indonesia had similar results. Surveying some of the world's last undamaged coral reefs, CI scientists found fifteen new species of corals and six new fish. One surprising discovery was the large number of fish which appeared to be exclusive ("endemic," in biological terms) to the area around the Togean Islands. As Dr. Gerald Allen of CI put it, "You rarely get this high degree of endemism in marine animals, especially among fish species."

Next came the results of a 1997 RAP expedition to Peru's Vilcabamba region. Dr. Louise Emmons, one of the world's leading experts on tropical mammals, was hiking through the mountain forest when she startled a weasel. The predator ran away, leaving its newly-killed prey like a gift at Emmons' feet. That prey was a new mammal, a two-pound rodent which had to be placed in a new genus. *Cuscomys ashaninka* is a gray-furred rat which may have an unusual history. Emmons believes it's related to a presumed-extinct mammal whose bones have been found in tombs in the Incan city of Macchu Picchu. The "Incan tomb rats" (now called *Cuscomys oblivata*) were apparently kept as pets and interred with their owners, much as cats were in ancient Egypt. Emmons speculates the tomb rat, too, may still be alive in the dense forests.

The same expedition turned up new butterflies, frogs, lizards, and two very interesting mice. The mice represented new species in the genus Juscelinomys. This was a genus previously known only from the type series (a group of nine specimens belonging to the type species). That type species was *Juscelinomys candango*, the Brasilia burrowing mouse—itself only discovered in 1960 and never caught since.

The work of the intrepid scientists on the RAP teams is always arduous and rarely safe. In August 1999, ichthyologist Fonchii Chang Matzunaga and boatman Reynaldo Sandoval drowned on an expedition in Peru. Mary Hagedorn, Matzunaga's colleague and friend, commented, "As romantic and adventuresome as it may seem, conservation is also extremely dangerous. Fonchii understood and had experienced this danger but was a woman who believed in her convictions and always followed her heart and dreams. She would not have changed one minute of her life."

This is the kind of dedication involved in cataloging the fauna of the world. Those of us who work safely at home can only support and salute people like Fonchii Matzunaga.

HOW MANY MAMMALS?

As we've seen, new mammals are still being discovered by people like Tim Flannery and the Conservation International teams. But how many undiscovered mammal species are still out there? Is it possible the number is in the thousands?

Yes, according to biologist Lawrence Heaney. Heaney, of Chicago's Field Museum, has described fifteen new species himself since 1991. He predicts the current catalogue of 4,600 mammal species may expand as high as 8,000. Seem ridiculous? Not when we've described, by Heaney's count, 459 new species between 1983 and 1993 alone. Most of these are rodents or insectivores, but they also include such finds as beaked whales, muntjac deer, a menagerie of primates, and several other sizable mammals we'll address later in this section. One of Heaney's own discoveries was fairly spectacular—a "flying fox" bat from the Philippines with a three-foot wingspan.

In 1984, no less an authority than George Gaylord Simpson, one of the most influential paleontologists of all time, wrote in a strongly worded dismissal of cryptozoology that scientists who sought new mammals were wasting their time. Only a relative handful of mammals could possibly remain undiscovered, and even those would be physically small and scientifically unimportant. It is thought-provoking to realize how dramatically wrong Dr. Simpson has been proven.

Not all new mammals are specimens brought in from distant islands and remote forests. About two-thirds of recently described species come from reclassification of existing types or identification of previously overlooked museum specimens, while the remaining third are being identified for the first time from the wild.

Several particularly important reclassifications have been announced in recent years. In 1998, a team led by researchers from the American Museum of Natural History announced that all the world's tigers were not, as long believed, variations of the single species *Panthera tigris*. The rare Sumatran tiger's DNA indicated this cat, which has developed for thousands of years in isolation on its Pacific island home, is a species distinct from all other tigers.

This is a good place to pause and mention what is being done with classification using DNA. DNA analysis is increasingly important as a tool for identifying species and determining their relationships. Several factors are taken into consideration when classifying living organisms. DNA markers are one indicator used to differentiate among families or genera and to draw lines between species belonging in the same genus. Within species, DNA is studied to determine how long individual populations have been reproductively isolated.

However, speciation based on DNA not a neat business. There is no universally accepted standard concerning the degree of difference between genetic sequences needed to define separate species. The amount of genome sequenced is an important factor, as is the number of specimens in the population sampled and the type of animal being studied. Therefore, differences in DNA are still considered only one of the factors used to identify species, along with known reproductive isolation (the basis of the dominant paradigm of classification, the biological species concept), and physical characteristics (morphology).

In late 1999, a team including Dr. Colin Groves reported the

African elephant was not one species, but two. Groves, a prominent anthropologist and taxonomist at the Australian National University, has a strong interest in identifying and classifying new species, especially mammals. (Not surprisingly, he is going to turn up several more times in this book.) According to Groves, skull characteristics and DNA analysis proved the forest elephant was a species, not merely a subspecies of the larger bush elephant (*Loxodonta africana*), even though the two types do interbreed in some areas. In 2001, four geneticists confirmed this view with additional DNA work.

A study published in late 2000 showed the North Pacific right whale was a full species, rather than a population of the northern right whale *Eubalaena glacialis*. In 2001, Chinese and British scientists announced that, based on DNA testing as well as observation, they may have found a new species of wild camel. The animal is outwardly similar to the two-humped Bactrian camel (*Camelus bactrianus*), but is uniquely adapted for what may be the harshest conditions preferred by any large mammal. The Kum Tagh sand dunes near Tibet offer only brackish water sources, and these camels survive on such salty water without any sign of the liver or kidney damage to be expected even in other camels. The estimated population is under 1,000.

Sometimes a specimen is in hand for a long time before anyone understands its significance. Over twenty years ago, Australian scientists captured several specimens of a small mouselike marsupial. These were classified as examples of a known species, the brown antechinus (*Antechinus stewartii*). Antechinus are fascinating because of their bizarre mating habits.

In 1998, curious researchers including Matthew Crowther of the University of Sydney took another look at the old specimens. Upon closer examination, they weren't quite the right size or color for brown antechinus. Skull measurements and tissue tests indicated the animals belonged to a distinct species. This discovery, christened the "agile antechinus" (*Antechinus agilis*), was the first new Australian marsupial to be named in a decade.

The next new Australian mammal was a somewhat similar case. In 1924, the purple-necked rock wallaby (*Petrogale purpureicollis*) was described. One of the features cited as unique was "purple pigmentation on the head and neck." Subsequently, this animal was relegated to the status of a subspecies whose holotype had been marked by "some kind of stain from the local foliage or rocks." Even when captive specimens developed the unique purple markings, the animal was considered to be a color variant of the black-footed rock wallaby. It wasn't until 2001 that a team of biologists determined, based on skull morphology and DNA analysis, that the strangely marked marsupial had been wrongly discounted.

Many more such discoveries are likely, according to Dr. Groves. In the summer of 2000, he and other experts from Australia and China began working to catalog a large collection of specimens which had been sitting overlooked for a hundred years in crates in the basements of two Chinese scientific institutions. The crates contained the collection of 19th-century naturalist and missionary Pierre Marie Heude. Groves found two specimens, a dwarf buffalo and a deer, which require investigation as possible new species.

It is the discoveries in the field, of course, which make the most news. An example of the finds still being made was announced in 1999. The story began when three very strange-looking rabbit skins with prominent black and brown striping across the face and all along the back were obtained at a market in Laos. This is a case where only the briefest glance was needed for scientists to suspect a new species. DNA analysis led by Diana Bell of the University of East Anglia confirmed that, and other scientists began combing Laotian forests, menageries, and marketplaces for live specimens. The Annamite rabbit, as it's now known, will join a similarly marked species from distant Sumatra in the genus Nesolagus.

An example of another kind of discovery came in 2002, when conservationists on the island of Borneo discovered a population of orangutans, estimated at 1,000 to 2,500 animals, whose existence had been unsuspected. While this find (which greatly cheered those worried about the fate of the species) involved a known animal, the fact that large populations of great apes could go undiscovered into the 21st century—demonstrated twice so far, once with these orangs and once with Ammann's chimpanzees—certainly implies there is a lot still to find out there.

Most of the discoveries concerning large mammals are described in separate essays later in this book. For now, I'll mention just a few mammals. University of California biologist James Patton found six new species (five rodents and an opossum) in just two weeks in Columbia. The same scientist led a team which found seven new species in Brazil in 1991. The Field Museum's Steven Goodman has found two new genera of rodents on Madagascar since 1991. The Panay cloudrunner, a chubby-faced two-pound rodent with thick brown fur and a long black tail, was described from the Philippines in 1993.

In 1997, a very small but newsworthy mammal discovery was

published. *Sorex yukonicus*, a tiny shrew from Alaska, was a rare example of a new mammal species from the United States.

The year 2002 brought, among other finds, a spiny rat from Brazil and another rodent with a long tail from Peru. Coming full circle back to museum discoveries, a dwarf mouse opossum from Argentina was described in the same year. The opossum, a rabbit-sized animal with a striking red underbelly, turned up as a single specimen in an old museum collection. Two new mammals described in 2005 are particularly noteworthy. The Australian snubfin dolphin, *Orcaella heinsohni*, was identified after a previously overlooked population of cetaceans with unusually small, rounded dorsal fins was examined in detail. From Tanzania came the first African monkey described in 21 years, the highland mangabey *Lophoccebus kipunji*.

During the years 1980-1990, more new mammals were described than in any decade since the 1920s. The "Golden Age" of zoology—often placed in the late 1800s—may actually be happening right now.

THE STILL-PUZZLING COELACANTH

There is no 20th-century animal discovery more famous than the coelacanth, *Latimeria chalumnae*. When this archaic-looking fish swam out of the depths of time and into a South African trawler's net in 1938, it proved descendants of a prehistoric species could exist undiscovered long after the end of the known fossil record. The credit goes to Marjorie Courtney-Latimer of the East London Museum, who first realized the fish was unique, and Professor James L. B. Smith, who realized the importance of what she had found.

It was fortunate Courtney-Latimer found the Mesozoic survivor at all. The specimen was buried under a ton or more of sharks and other ordinary fish on the deck of the trawler *Nerine*. The young museum curator noticed a single blue fin sticking out of the pile, and she dug through the mess to pull out the fish it was attached to.

That was on December 22, 1938. It was February 16 of the next year before Professor Smith, a prominent South African chemist and ichthyologist, could make it to East London to see the now-mounted specimen. Courtney-Latimer remembered, "He circled the fish several times in silence, went close to it, and stroked it. Then he turned to me and said, 'Lass, this discovery will be on the lips of every scientist in the world.'"

Smith was correct, because he had identified the first known survivor of the ancient order Coelacanthini. This fish has enough

The coelacanth, Latimeria chalumnae.

peculiar features to attract considerable scientific interest even without the "living fossil" angle. The most obvious are the pectoral fins, which look like stumpy, fringed legs. (The crew of the *Nerine* called it "the great sea lizard.") While the coelacanth is not our direct ancestor, it is a member of the lobe-finned fishes, the group which gave rise to the amphibians, which in turn evolved into the reptiles, mammals, and birds. The coelacanth has an oil-filled notochord instead of a bony spine and a unique intracranial joint in the skull that lets it open its mouth unusually wide. The fish's habits are as peculiar as its anatomy. It has been filmed hovering vertically with its head just off the bottom, possibly seeking the electrical signals of prey species. During the day, coelacanths often rest in underwater caves, with up to fourteen individuals peacefully sharing one shelter.

For sixty years after its discovery, this fish's habitat was presumed limited to the area immediately surrounding the Comoro Islands. This was true even though the very first coelacanth was caught 1,800 miles away, off South Africa. No more South African specimens were snagged, but there was a 1991 catch off Mozambique. Most experts classified both of these fish as strays. After additional catches near Madagascar in 1995 and 1997, a few researchers began to suspect there was a second population, closer to the African coast. Another coelacanth was snagged off Madagascar in 2001. While the evidence is becoming highly suggestive, a resident population off Madagascar has still not been located.

An icthyological bombshell exploded in 1998 when a second population was finally discovered—not off Africa, but thousands of miles east of the Comoros. The story began in September 1997, when Dr. Mark Erdmann, a biologist, and his wife Arnaz, a naturalist, were on honeymoon in Indonesia. Arnaz was walking through a fish market in Manado, on the island of Sulawesi, when she saw something odd on a fisherman's cart. "Isn't that one of those fossil fishes?" she asked. It was. The Erdmanns photographed the coelacanth and, on their return to the U.S., obtained financing from the National Geographic Society and other institutions to investigate.

On July 10, 1998, fisherman Om Lameh Sonathon hauled up a live specimen of the fish locally called *rajah laut*—"king of the sea" —off Sulawesi. He brought the coelacanth to Dr. Erdmann. Erdmann photographed it while still alive, and, when it expired after three hours, froze the fish to preserve it for examination. The Indonesian specimen was similar in morphology to *L. chalumnae*, with one minor but intriguing difference. Ornamental scales scattered over the body of Erdmann's specimen body provided a prismatic reflecting effect, making the fish look as through it were flecked with gold.

In 1999, two research teams, one led by Erdmann and the other by Laurent Pouyard of the Indonesian office of the French Scientific Research Institute for Development, analyzed the genetic makeup and morphology of the Sulawesi specimen. They agreed the Indonesian fish represented, not just a different population, but a different species than the known coelacanth. The new coelacanth was named *Latimeria menadoensis*. This name was bestowed by Pouyard and his Indonesian co-authors, who published their paper on the subject before Erdmann's was ready. Erdmann was understandably exasperated by this breach of professional courtesy. Unfortunately, the rules of taxonomic nomenclature recognize the name given in the first publication, without regard to the discoverer.

L. menadoensis appears to have a larger range than its Comoran cousin. A search using a submersible and led by Dr. Hans Fricke, who had succeeded in filming *L. chalumnae*, did not find coelacanths where Erdmann's specimen was caught. However, they later observed two coelacanths at a spot some 200 miles to the southwest.

Regardless of the exact range of each species, it is obvious the 6000-mile distance between Sulawesi and the Comoros is too great to support any "stray" explanation for the new coelacanth. Having two completely isolated populations so far apart is extremely unlikely. One ichthyologist, Dr. John McCosker, estimated that anywhere from two to nine million years of evolution separate the two species. Pouyaud's team put the gap at 1.2 to 1.4 million years. Either way, it's taken a long time to develop two widely separated populations of a slow-moving fish. The odds are that whatever migration took place has left other coelacanth communities in between these two areas—and maybe beyond.

There is a bizarre postscript to this story. Soon after the reports of the Indonesian specimen were published, French cryptozoologist Michel Raynal received an intriguing message from Georges Serre, a fishery specialist with a French research institute. Serre claimed to have caught a coelacanth near Java, over 900 miles from Sulawesi, in 1995. Serre said he had photographed the fish but lost the picture, and the specimen vanished during shipment.

In the spring of 2000, Serre relocated his picture and submitted it to *Nature* for publication. Unfortunately, the staff at *Nature* smelled a rat—or a dead fish. The specimen in Serre's photograph looked identical to the coelacanth obtained by Erdmann. Serre, while not admitting any fraud, changed his story, saying the photograph was not taken by him, but by a friend who had since died. At this writing, the tale of the alleged Javan coelacanth ends here.

Late in 2000, the idea of a population off South Africa was dramatically confirmed. Diver Pieter Venter was 320 feet below the surface off that country's northeastern coast when he saw three fish which were unmistakably coelacanths. Venter gathered a team and went back with cameras. The depth at which Venter had his encounter is pushing the limit of safety for free divers, and one of Venter's assistants died after surfacing too quickly. Film obtained on the dive showed three coelacanths, which looked very much like the ones from the Comoros. Since then, the government has announced a major coelacanth research program, and entrepreneurs in nearby Sodwana Bay are selling coelacanth T-shirts and otherwise promoting the animals as "like our own Loch Ness Monster."

Marjorie Courtenay-Latimer, after a long life filled with scientific and educational work and honors, died on May 17th, 2004. She had lived to see the scientific world learn much more about her strange fish, as well as to ponder the mysteries yet to be solved.

One of those remaining mysteries centers on the tantalizing hints of other coelacanth habitats.

Four coelacanths have been caught of Tanzania in 2003 and 2004. These, combined with the Mozambican catches, raise the possibility that this seemingly sedentary fish often wanders further from its home ranges than thought, or that there could be a chain of coelacanth habitats off East Africa which we have yet to pinpoint.

Two small silver votive figures of coelacanths found in Spain have been offered as proof by researchers who believed they were made in Mexico at least a century before. In 2001, however, a new analysis described these figures as modern Spanish artworks based on known coelacanth specimens.

That still leaves us with a puzzling event which took place in 1949. In that year, a Tampa, Florida, souvenir shop bought some peculiar scales from a local fisherman. The shopkeeper sent one to Dr. Isaac Ginsberg of the U.S. National Museum, who thought it came from a coelacanth or a close relative. Unfortunately, Dr. Ginsberg never heard from the shop owner again. Even the current whereabouts of the sample scale (which one coelacanth specialist, Dr. George Brown, suggests was misidentified) are unknown.

In 1995, reports circulated about a coelacanth being caught off Jamaica. This startling tale made some newspapers, but no one was able to confirm it. The aforementioned George Brown, founder of a coelacanth conservation group, the Society for Protection Of Old Fishes (yes, the acronym is SPOOF), reported he'd heard nothing about it. Dr. Karl Shuker, an English zoologist who also tried to verify the story, believes it was a hoax or a case of mistaken identity.

There rests the mystery of coelacanth distribution. It's likely there are (or were) populations between the Comoros and Indonesia. It's also still possible we'll find coelacanths in other parts of the globe. Perhaps this famous fish, while still fascinating and important, isn't nearly as rare as we thought.

THE MAMMALS OF VU QUANG

The string of recent discoveries involving large hoofed mammals in Southeast Asia is continuing, seemingly with no end in sight. A series of expeditions, sponsored mainly by the Wildlife Conservation Society (WCS) and the WWF (a.k.a. the Worldwide Fund for Nature or the World Wildlife Fund) has uncovered one new or "extinct" species after another since 1992.

The focus of these discoveries is the Annamite Mountains, a broken range of rugged, forested slopes and peaks running roughly southeast to northwest for about 350 miles. Along most of its length, this natural barrier forms the border between Vietnam and Laos. About two thirds of the way up the range, where the territory of Vietnam measures less than fifty miles across from Laos to the South China Sea, lies the province of Ha Tinh, location of the now-famous Vu Quang Nature Reserve.

This reserve constitutes an area so difficult and remote it was almost totally bypassed during the wars that swept through this region. The animals that found relative safety here remained unknown to zoology until the 1990s, when a series of discoveries in Vu Quang shattered the belief that the major land mammals of the world had long since been catalogued.

Many of these discoveries resulted from the classic method of asking the inhabitants of remote regions about their local fauna and following up to search for hard evidence. While none of the scientists involved may call themselves cryptozoologists, their spectacular finds validated the basic premise of cryptozoology: that many animals, including some surprising and significant ones, still await discovery. There have been small mammals discovered in this region, such as a new squirrel in Vietnam's Gia Lai province and the gray-shanked douc langur (*Pygathrix cinereus*) in Ninh Binh province, but it is the large mammals which earned the area its permanent fame in the zoological world.

At this writing, the score on such mammals in Southeast Asia over the last fourteen years stands like this:

SAOLA (VU QUANG OX):

The animal formally described in 1993 as *Pseudoryx nghetinhensis* is easily the zoological find of the half-century. In May 1992, Dr. John MacKinnon led an expedition to the Vu Quang region. Vietnamese zoologist Do Tuoc, after interviewing local hunters, turned up several skulls with long, almost straight horns. These were being kept as trophies in the homes of the area's few residents. He showed the skulls to MacKinnon, who immediately recognized them as belonging to a new animal but was thoroughly puzzled about what it might be. He thought it could be related to the anoa, a dwarf buffalo. When skins of the whole animal were collected, though, it looked more like an antelope. As it turned out, the creature was neither buffalo nor antelope, but something far more unusual.

Briefly called the Vu Quang oryx, the animal Vietnamese call *saola* ("spindle horn") was eventually given its genus name, which means "false oryx." It appears to be the sole living species from a heretofore-unknown branch of the bovid family tree. The saola is a striking animal by any name. It is predominantly brown with prominent black and white markings, especially on the head. A large adult saola may weigh over 200 pounds. The total population of this species is estimated in the hundreds. The first living specimen, a juvenile, was caught in June 1994. Unfortunately, this and all subsequent specimens kept in captivity have died after a few weeks or months.

The giant muntjac of Laos, described in 1996.

GIANT MUNTJAC:
Also called the Mang Lon muntjac, this species was described in 1996 as *Megamuntiacus vuquangensis*. Rob Timmins and Tom Evans, two British ornithologists, began the search for this animal when they noticed some unusually large muntjac horns displayed in Laotian homes in the Nakai Nam Theun reserve (adjacent to the Vu Quang reserve across the border in Vietnam). Laotian hunters described the source of these horns as a unique type of muntjac, larger than the red muntjac (*Muntiacus muntjac*) they hunted most often. The expedition members collected skulls and antlers of the unusual deer, then happened across a live specimen in a menagerie in Lak Xao. By this time, MacKinnon had also collected a skull of this 100-lb animal in Vu Quang. It was he who first called it the "giant muntjac." George Schaller and Elisabeth Verba wrote the formal description. The muntjac's status in a separate genus is disputed by some experts, but all agree it's a very distinct species.

TRUONG SON MUNTJAC:
Muntiacus truongsonensis was described by John MacKinnon and others in 1998. One of the smaller muntjacs, it weighs about thirty-three pounds. Its remains were collected in April 1997 by WWF scientists and their Vietnamese counterparts, including Do Tuoc.

The first zoologist who saw a skull in a hunter's home thought the very small trophy was from was a juvenile muntjac, but Do Tuoc took a closer look and noticed the skull had adult teeth. Eighteen similar skulls were obtained from local hunters. It was established that the new species lives in the Truong Son mountain range in western Vietnam, in the province of Quang Nam. The Truong Son muntjac's horns lack the brow tine common on the horns of other muntjacs. Both sexes have long canine teeth, which in most muntjac species are borne only by males.

LEAF MUNTJAC:

Muntiacus putaoensis was described by George Amato, Alan Rabinowitz, and Mary G. Egan in 1998. Its story began in the spring of 1997, when Dr. Rabinowitz encountered the animal in northern Myanmar. To explore the region which turned out to harbor the new muntjac, Rabinowitz and several colleagues spent two months tramping 250 miles into very high and very cold mountain ranges where no scientific survey had ever taken place. Most of the team came down with pneumonia, and Rabinowitz could hardly walk due to knee problems. Rabinowitz collected several deer heads kept by hunters as trophies, but didn't think at first that they represented a new species. Then the visitors met a local hunter who displayed a complete, freshly-killed specimen. As Rabinowitz recalled it, his response was, "Hey, this is something totally different."

This chestnut-colored mammal weighs only about twenty-five pounds and stands less than two feet high. The common name derives from the fact hunters can carry its body on a large type of leaf.

THE MYSTERY GOAT:

A final and most bizarre episode from this region concerns a large animal described in 1994 as Pseudonovibos spiralis. The description was based entirely on the animal's peculiar spiral horns, which are about eighteen inches long and twist dramatically, so they resemble high-rise motorcycle handlebars. Biologist Wolfgang Peter found the first set of horns in a Ho Chi Minh City market stall. Despite the paucity of evidence, the horns were so different from those of any known animal that Peter felt justified in describing a new species and genus. According to Peter and other researchers, the animal's range apparently extended into Cambodia, where it was called kting voar (or khting vor), meaning "jungle sheep." Cambodian hunters described the animal as gray or dark reddish in color. The Vietnamese name, linh dong, was most commonly translated as "mountain goat."

Subsequent expeditions turned up a few more specimens, including frontlets (skull frontal bones) with horns attached. A few older examples also turned up, having been overlooked or misidentified in museum collections, and it was discovered that a Chinese text dated 1607 showed an unidentified goatlike animal with similar horns. Conservationist Hunter Weiler, who works at the Herculean task of saving the mammals of Southeast Asia, commented, "To my knowledge, this is the only wild mammal on the face of the Earth that we positively know exists but have never seen."

The story, though, had a twist even more dramatic than the animal's horns. In December 2000, three French scientists who had examined some of the evidence declared the "goat" was a fake. According to Arnoult Seveau of the Zoological Society of Paris and his colleagues, the horns had been manufactured from those of cattle. The horns had been heated and bent into the desired shape, and the spiral rings added by whittling them out with a knife. "*Pseudonovibos spiralis* is simply a forgery," the trio concluded. The following year, Seveau and five French and Vietnamese colleagues published results of DNA tests that reinforced the claim.

Defenders responded that some horns were nevertheless authentic. Professor Robert Timm of the University of Kansas, who has studied the animal's taxonomic relationships, responded, "There is NO doubt in my mind that *P. spiralis* is a real animal, a valid taxon. The specimens we have here are very clear." He and John Brandt published a paper on two previously-unidentified frontlets collected in 1929, arguing these showed no signs of having been reworked from cattle horns and could not be matched to any known species. Subsequently, a group of Russian scientists led by G. V. Kuznetsov weighed in with a published analysis of mitochondrial DNA samples taken from other specimens of this controversial animal. They found the specimens they examined could not be cattle remains. Instead, the animal was a distinct species most likely aligned with the buffalo (that, is the genus Bubalus, often called water buffalo), making it quite distinct from the domestic cow (*Bos taurus*).

While some of the definitely faked *P. spiralis* horns are decades old, others are recent. It's likely that, after Peter's 1994 discovery, Vietnamese and Cambodian entrepreneurs dug out any old fakes they could find and made some new ones to foist (successfully) on visiting zoologists. Just when manufacturing the horns of this animal became a cottage industry, and why, is unknown. Indeed, the khting vor lives in a forest of questions, as much as in a rain forest. Is there a real species? If so, is it living or recently extinct? If it's still existing, where is it? It is possible that sightings of other animals, like the kouprey and guar, combined with the region's confusion of languages, dialects, and traditions (Laos alone has 240 ethnic groups) has so obscured the origins of the khting vor mystery that the truth will take a long time to emerge.

Whatever the final status of *P. spiralis*, the bonanza of genuine large animal discoveries from one area remains unlike anything that has happened in zoology since the 19th century. If so many mammals could be lurking in one place, there is every reason to believe we have missed others—in this and other regions of the world.

FINDING A PHYLUM

In 1995, a new phylum, Cycliophora, was added to the animal kingdom. There are only thirty-some accepted phyla, and an animal has to be very unusual to merit placement in a new one. In proposing a new phylum, a researcher is saying that he or she has found an animal which is not only a new species, but cannot be placed in any genus, family, order, class, or phylum of known animals.

The species involved in this case, *Symbion pandora*, passed this imposing test without much difficulty. Symbion is about the size of the dot on the letter i. It was discovered three decades ago on the mouthparts of a lobster, but the initial catalogers placed it in a new genus and left it at that. A reexamination by Peter Funch and Reinhardt Kristensen of Copenhagen University showed Symbion resembled nothing else on Earth. Dr. Simon Morris of the University of Cambridge called Symbion "the zoological highlight of the decade."

The phylum name is Greek for "carrying a small wheel" and refers to the animal's round, cilia-fringed mouth. Most of the press and public fascination has centered around the animal's strange reproductive habits. At different stages of its life, Symbion reproduces sexually by "budding off" male and female offspring, or asexually, in which case its digestive tract metamorphoses into a larva. Even American humorist Dave Barry paid attention to this story, writing, "Zoologists, who don't get out much, are excited over an animal that basically reproduces by pooping."

In creating Cycliophora, Funch and Kristensen erected the first new phylum since Vestimentifera was proposed to house the giant tube worm, *Riftia pachyptila*, in 1985. Kristensen believes there are countless discoveries still to be made concerning tiny marine life. "This is only the beginning," he said. "When we have finished, the zoological system will be turned upside down."

Kristensen had his reasons for making such a bold statement. In 1983, he had named another new phylum, Loricifera, for miniature animals which look sort of like animated pineapples with snouts.

Loriciferans were first identified in 1974 from seafloor samples taken off the coast of France. They burrow through gravel or sand on the floor of shallow areas of the ocean. The phylum name means "girdle wearer," in reference to the ring of scalelike structures which encircle and protect the animal. The snouts are mouthparts which can be retracted into the body: indeed, a Loriciferan can retract its entire head. At least ten species have been collected from depths of fifty feet to 1,500 feet off Europe and North America.

In October 2000, Kristensen and Funch proposed yet another phylum. Micrognathozoa was named to house *Limnognathia maerski*, a miniscule creature found in the frigid fresh water of wells in Greenland. Only 1/250th of an inch long, it nevertheless has unusual, complex jaws used to scrape algae from rocks. The species appears to consist entirely of parthenogenic females.

Identifying four new phyla in the last two decades shows how much we still have to learn about life of a size we rarely think about.

OF LEMURS AND POTTOS

The recent discoveries of a new orangutan population and Ammann's odd chimpanzees created quite a buzz in the scientific community. One reason was that they were exceptions to the rule that most newly discovered primates are small animals. Examples of these include several recently described marmosets and other monkeys from South America, which are discussed later in this section.

Another example is the golden-brown mouse lemur, *Microcebus ravelobensis*. This primate was described in 1998 from forests in northwestern Madagascar's Ampijoroa Nature Reserve. A good-sized specimen may be ten inches long, tail included. The new species has long, naked ears and an impressive ability to jump between branches.

This isn't the only recent lemur discovery. All known lemurs are endemic to Madagascar or the Comoros Islands, so one would think they would have been cataloged many years ago. Not so.

The golden bamboo lemur (*Hapalemur aureus*) was described in 1987. It's quite a bit larger than *M. ravelobensis*, weighing between three and four pounds. Primatologist Patricia Wright first spotted this animal in Madagascar's Ranomafana National Park in 1985. In 1988, the golden-crowned sifaka (*Propithecus tattersalli*) was described from Madagascar's northern tip. This animal, also known as Tattersall's sifaka for its discoverer, Ian Tattersall, is a rich yellow-orange color and is robust for a lemur, weighing up to eight pounds.

In November 2000, an international scientific team reported that another three new species of mouse lemurs had been discovered, with two more awaiting confirmation. All were chipmunk-sized animals inhabiting densely forested regions of Madagascar's west coast. One researcher involved, Steven Goodman of the Field Museum, commented, "It's incredibly rare to discover a new species of primate, let alone three new species." One of the new species was christened Berthe's Mouse Lemur (*Microcebus berthae*). Looking like a golden hamster with big eyes and a long tail, it claimed, at least temporarily, the title of "world's smallest primate."

In 2002, another contender for that title turned up. Researchers Mireya Mayor and Ed Louis discovered another tiny new mouse lemur in a live trap they'd set as part of a survey of lemur populations. Mayor, an anthropologist whose unique background is always mentioned in press accounts—she is a Fulbright scholar and a former cheerleader for the Miami Dolphins professional football team—explained the significance of the new find this way. "You have this new discovery, this new animal, and people say, 'Wow, there's so much we don't know.' And people will be drawn to this particular area because of this; it can raise awareness and get the government involved because there's something new and exciting to focus on. You can say, 'Look at what we just discovered.'"

In 2004, Wright told an interviewer she was pursuing another new species, a large lemur with big ears and reddish-brown hair. Two more lemurs were described by German scientists in August 2005.

In 1996, another kind of small primate fell into the lap of science: not from a tree, but from a drawer. When anthropologist Jeff Schwartz opened an old specimen drawer at the University of Zurich, he found it held two skeletons labeled as a known species of potto. Pottos are African primates, usually about ten inches long. Along with the lemurs, lorises, and tarsiers, they belong to the suborder called the prosimians, or, to use the more colorful German term, *halbaffen* (half-monkeys).

Schwartz's twenty years spent studying pottos and their relatives immediately told him something was wrong with the labeling of these two specimens. "Pottos have very short tails; this one had a longer tail," he observed. "Pottos have big spines sticking out of the neck vertebrae; this thing didn't. This creature's teeth didn't have any of the salient features of potto teeth. Also, it was the smallest size of anything that's been called a potto, but it was collected in an area that's home to the world's largest pottos."

After closer study, Schwartz identified a new guinea-pig-sized animal that belonged in its own genus. Since the new species was clearly related to the pottos, he named the genus Pseudopotto and the species *Pseudopotto martini*. Schwartz has suggested the animal is so unique it might even merit naming a new family.

Could there still be living Pseudopottos? No one knows, but Schwartz commented, "It is very exciting to think that, somewhere in the tropical forests of Cameroon, Pseudopotto lives."

THE SOCIAL SHRIMP

A recently discovered species of shrimp, *Synalpheus regalis*, has caught the attention of science because of its unusual lifestyle. The half-inch-long animals, described in 1996, live in colonies of up to 300 individuals located inside sea sponges in the Caribbean. The colonies are eusocial: that is, like the social insects, they have an ongoing society in which overlapping generations cooperate in raising the young, defending the colony, and so forth, and there is a single breeding queen. These shrimp were the first known eusocial marine species of any type.

The describer was J. Emmett Duffy of the Virginia Institute of Marine Science. In studying colonial shrimp, Duffy was initially puzzled by the lack of females. He plunged into further examination of the colonies and the individuals, attempting to unearth the reason for this oddity. Eventually, he found that each colony was "essentially a two-parent family with a whole lot of grown male children hanging around." Most, if not all, of the shrimp in each colony are siblings, offspring of the current queen. Juvenile members of the colony develop inside the "natal sponge." Like most eusocial species, the shrimp don't like strangers. When members of a related species were introduced into eight colonies, all eight intruders were killed.

In 1998, Duffy co-described another new species, *Synalpheus chacei*, which is also eusocial and lives in sponges. He added a third species of sponge-dwelling shrimp, *S. williamsi*, in 1999, although this species did not display eusociality. Several of the known colonial species, on closer examination, have turned out to be living the social-insect pattern of life.

In 1999, a very different shrimp was discovered in Australia's Sydney Harbor, a place thought far too well-known to shelter any new animals. *Erugosquilla grahami* is one of the mantis shrimps. This group is so named because their lethal pinchers, normally held in a folded position, resemble those of the praying mantis. The new purple-and-blue shrimp is a robust species, about eight inches long. *E. grahami* may have the fastest strike of any animal known. It can grab its prey in five to eight milliseconds. To find and seize that prey, it has the most complex eyes of any invertebrate. The shrimp's eyes are highly specialized, with each eye independently capable of binocular vision.

Biologists are eager to make further studies of the creature's unusual vision and the neural equipment required to coordinate its complex senses and lighting-fast strikes. Other Australians, not surprisingly, have their own ideas on what to do with the shrimp. Shane Ahyong, the species' discoverer, reports, "They're quite nice if they're barbecued or char-grilled—but not overcooked."

SOUTH AMERICA'S NEW BIRDS

South America is home to over 3,000 species of birds, almost a third of the always-changing total believed to inhabit the Earth. No other continent has such a diverse avifauna. Not surprisingly, then, South America remains a hotbed for encounters with previously-unclassified bird species. By one expert's count, ornithologists discover an average of four new birds a year in South America.

Let's look at some interesting recent examples.

The discovery of one unique bird, requiring creation of a new genus, was announced from Brazil in 1996. This find came from a surprising location. *Acrobatornis fonsecai*, a diminutive black-and-gray member of the oven bird family, was found living in large nests which were clearly visible from the main highway in Brazil's state of Bahia. No one knows why it wasn't found long ago.

The species was finally brought to the attention of science by a dedicated amateur birder, Paulo Fonseca. Ornithologist Jose Pacheco said, "Imagine how many biologists drove by and never even noticed." The bird, incidentally, runs along the undersides of branches, spending more time upside down than right side up. This is the origin of the genus name, which means "acrobat."

In November 1997, in Ecuador, American ornithologist Robert Ridgely and four colleagues were recording bird songs in the Andes when they heard a most unusual call, described as "akin to an owl's hoot and a dog's bark." The bird responsible was eventually tracked down and caught, and it was described in 1999 as a new species of antpitta, *Grallaria ridgelyi*. (Dr. Ridgely did not write the formal description: the ornithologists who did opted to name it after him.) Ridgely learned the bird was known by the local name of *jocotoco*.

The jocotoco is about the size of a quail, with long, blue legs and very limited flying abilities. Other distinguishing marks include very short tail feathers and a distinctive white facial stripe. Ridgely described the new species this way: "It's a pretty snazzy bird. It looks like a big, round fluff-ball on pogo-stick legs."

The jocotoco lives near the Podocarpus National Park in a "well-explored" area where no one, including Ridgely, thought there were new species to be found. Shortly after the discovery was announced, British and American conservationists formed the Jocotoco Foundation to purchase land in the bird's habitat and maintain it as a refuge. On finding the bird in an area where nothing new was expected, Ridgely commented, "It's like finding a pot of gold in your backyard. It certainly makes me wonder what else is out there if something this unusual could go undetected for so long."

The new antpitta came on the heels of another Brazilian find, a very small, dark-feathered bird christened the lowland tapaculo (placed in the genus Scytalopus). The tapaculo, too, was discovered by ornithologists after they heard calls they couldn't identify. The tiny, black-and-gray bird was captured in a marsh within the city of Curitiba in 1997 and announced a year later, after its identity as a new species was confirmed. Actually, tapaculos seem to be turning up a lot: one new species was described from Bolivia in 1994, and three were added from Ecuador in 1997.

In 1998, the ancient antwren (*Herpsilochmus gentryi*) was described from Peru. As with so many recent discoveries, the first clue to this bird's existence was a vocal one. Over the course of several years' work in northern Peru, ornithologists Jose Alvarez Alonzo and Bret Whitney recorded the vocalizations of the region's avifauna. In 1994, Whitney was listening to some of the tapes when he picked out a strange call. It seemed to belong to an antwren in the genus Herpsilochmus, but did not match a known species. In January 1995, Whitney and Alvarez mounted an expedition up the Rio Tigre in search of the mystery bird. Not only did they find it, but investigations by another ornithologist demonstrated that the new species' range extended into Ecuador.

In 1999, the Andes Mountains straddling Ecuador and Columbia produced the cloud forest pygmy owl. *Glaucidium nubicola* was described by Mark Robbins of the University of Kansas Natural History Museum and Gary Stiles of the National University of Columbia in Bogota. The new species inhabits only the forests above 4,500 feet on the Pacific slopes of the Andes. This marks the fourth bird species Robbins has described in the last ten years. In 1995, for example, he co-authored the description of *Glaucidium parkeri*, the subtropical pygmy owl, which ranges over portions of Ecuador, Peru, and Bolivia.

The cloud forest pygmy owl is a small, shy bird, and was only identified when careful study of museum specimens and recorded calls led Robbins and Stiles to the discovery that type specimens of two new species (the other being the Costa Rican pygmy owl, *Glaucidium costaricanum*) had been misclassified as specimens of the known Andean pygmy owl. One specimen of this brown and white bird had been collected in 1987, the other in 1992. Another pygmy owl, the two-ounce *Glaucidium mooreorum*, was described in 2003. This critically endangered bird was named for Gordon Moore, founder of Intel Corporation, and his wife Betty. Naming a species for a businessman and his wife rather than a discoverer is a bit unusual, but in this case was appropriate, as the Moore Foundation had given Conservation International a record $261 million grant for conservation programs.

In 2000, a description was published of a distinctive new species from the Eastern Andes of Peru. The scarlet-banded barbet is a handsome bird, with scarlet feathers on the crown and nape plus a wide scarlet band bordering the white throat and upper breast. The song of this new species, *Capito wallacei*, is described as a "hollow purr." Four specimens were netted in a July 1996 expedition to an unnamed mountain peak on the east side of the upper Rio Cushabatay.

More recently, ornithologist Andrew Whittaker also heard a strange sound. In Whittaker's case, he was in the Brazilian state of Para, watching and videotaping the bird life. He didn't immediately spot the bird providing the unfamiliar call, but found it by playing back a videotape. It was a new falcon, which Whittaker in 2002 named the cryptic forest-falcon, *Micrastur mintoni*.

The paper describing the scarlet-banded barbet included the intriguing note, "The members of the 1996 expedition encountered several other 'mystery' birds that were not collected and likely are undescribed taxa." Events since then are continuing to bear out the importance of scouring the jungles and trees of South America.

THE SMALLEST FROG

A diminutive but nonetheless interesting herpetological discovery from Cuba was announced in 1996. The subject of this excitement was the Northern Hemisphere's smallest frog. In fact, the "eleuth frog" (so called after its genus, Eleutherodactylus) is the hemisphere's smallest known terrestrial vertebrate of any kind.

Under half an inch long and about the size of a fingernail, the dark brown frog with a copper stripe down its back is only a hair too large to claim the title of smallest terrestrial vertebrate in the world. That honor goes to a Brazilian frog, itself discovered only in 1971. Dr. S. Blair Hedges, one of the eleuth frog's discoverers, commented, "It's not one of those things where you go out and search for the largest or smallest. You just sort of stumble onto it."

A much larger and far stranger frog was found in 2003. *Nasikabatrachus sahyadrensis* is a weird-looking creature from the mountains of southern India. This frog has been compared in appearance to a half-squashed plum. It is a dark purple animal, about three inches long, with a small head and pointed snout. No frog family had been created since 1926, but this creature definitely required one. Its nearest relatives live in the Seychelles, near Madagascar, separated from it by 2,000 miles and 65 million years of continental drift.

Are other new frogs and related animals still being discovered? Yes—to say the least. Consider the list of discoveries herpetologist Chris Raxworthy has piled up in Madagascar. As of 1997, he and his colleague, Ronald Nussbaum, had already catalogued dozens of new herps (reptiles and amphibians). He had a backlog of species awaiting description which included numerous lizards and approximately one hundred frogs.

In 2003, Raxworthy and his colleagues used specimen locality data and satellite imagery in a project assisted by the U.S. National Aeronautics and Space Administration (NASA) to map the habitats of Madagascar's chameleons. The resulting data maps showed something not intended—areas of "error" where the predicted distribution did not match the specimen data. When herpetologists checked these areas in person, they found that seven previously unknown species had been occupying those spaces, thus throwing off the predictions concerning known species. A promising new tool had been discovered along with the seven chameleons—species soon to be added to the impressive 942 herp species formally described worldwide from 1990 through 1997.

Closer to home, thirty new species of salamanders were discovered between 1943 and 1998 in the United States and Canada alone. One of these, found in 1997, turned up in the San Gabriel Mountains only twenty-five miles from Los Angeles. Other recent herp discoveries from the U.S. include *Rana subaquavocalis*, a new species of leopard frog. Biologist James Platz collected the type specimen in Arizona's Huachuca Mountains in 1990. *R. subaquavocalis* was so named because it's the only frog in the world to emit its mating call underwater.

Discoveries in herpetology rarely make the news. In the case of snakes, for example, a new species is likely to draw attention only if the serpent is large, poisonous, or both. An example of the poisonous type is the Burmese spitting cobra, *Naja mandalayensis*. Described in June 2000, it is smaller and darker than other spitting cobras and lives in the dry central regions of Myanmar. Reports of this animal go back for a hundred years, but it had never been identified until herpetologist Joseph Slowinski collected the type specimen in 1998. This snake, fairly small for a cobra (the type specimen was thirty-three inches long) can spit its venom accurately for over six feet.

Some other venomous snakes recently added to the world's reptiles include finds from Mexico and Australia. The Mexican discovery is a new coral snake from the country's high deserts. Described in 2000, *Micrurus pachecogili* is marked with bright yellow rings and short red rings on its black body. From Australia in 1998 came the highly poisonous false king brown snake, *Pailsus pailsei*. Brown with a "slight olive tinge" on top and with a creamy yellow underside, this snake may be up to six feet long. It is so far known only from the area near Mount Isa in Queensland.

Finally, we return to the amphibians to note how British naturalist Martin Pickersgill has reminded the scientific world of the value dedicated amateurs still offer to zoology. Pickersgill, who discovered a new frog species in 1983, more recently completed a ten-month solo expedition through Africa. The specimens and observations gathered on his trip resulted in four more new species of frogs being identified.

THE CARNIVOROUS SPONGE

Sponges, as a group, are about the most inoffensive animals imaginable. After all, they spend their entire lives clinging to rocks or corals, passively filtering tiny bits of organic debris from the water. Sponges are among the most primitive multicelled animals on Earth. One invertebrate biologist, Sally Leys, called them "the blobs of the animal world." Sponges don't even have nervous systems or any sort of digestive tract.

There may be 10,000 species of sponges, and scientists are still finding new ones. Sponges are not as boring a group as they may appear. Some species may live for hundreds of years. At least a few species move—a few millimeters a day, anyway. This is quite puzzling for an animal in which everything takes place at the cellular level, and all the cells have to move in concert (with no brain to direct them) to get anything done.

While some sponges are big enough for a diver to crawl into, the most amazing discovery from the phylum Porifera is a tiny one.

This surprise from the sponge world was announced from the Mediterranean Sea in 1995. There biologists discovered a new species (*Asbestopluma hypogea*) which had abandoned filter-feeding and adapted to catching and eating tiny crustaceans. Hook-shaped spicules of silicon, resembling Velcro, cover long filaments extending from the animal. Prey animals are entangled in these filaments and held as additional filaments grow over them and, over the course of a few days, consume them. To do this, individual cells migrate, changing their positions in the organism as soon as food is trapped. Marine biologist Jean Vacelet explained, "The sponge is like a giant amoeba."

This diminutive sponge, under two-thirds of an inch high, offers several puzzles. It has abandoned the basic body plan, optimized for flushing water through the system, which characterizes its phylum. In addition, its closest relatives are deep-ocean sponges living at depths of up to 29,000 feet. What is it doing in a Mediterranean cave only fifty feet down? Even though the cave's dark, cold water and limited water circulation make the environment similar to much deeper areas, how did the ancestors of this species get there? And while feeding on larger prey (macrophagy) is a logical adaptation to areas where the flow of nutrient particles is inadequate, how did the sponge change so drastically?

Sponge experts are still investigating these questions. Vacelet and his co-discoverer, N. Boury-Esnault, state, "Such a unique body plan would deserve recognition as a distinct phylum, if these animals were not so evidently close relatives of Porifera." A second carnivorous species, found off the far away Aleutians in 2005, now awaits a name.

THE UNKNOWN HORSES

The horse is probably the second animal humans learned to tame. Our association with these handsome and useful beasts goes back thousands of years. In that time, breeders have scoured the world for animals with desirable traits, and selective breeding has produced highly specialized types for racing, farming, pleasure riding, and other equine occupations.

Despite this familiarity, two entirely new breeds have been discovered in the last few years. A new type of horse is not a new species, and probably does not even merit subspecific rank. Still, these are cases where herds of large mammals, living in association with humans, have existed until recently outside the knowledge of science.

Dr. Michel Peissel, a French explorer and anthropologist, discovered the Nangchen horse in Tibet in 1993. Peissel had been traveling through Tibet studying the breeding and use of horses in general, while looking specifically for an isolated breed mentioned in ancient Chinese records. When he found it, he was highly impressed. The Nangchen horse is a large, powerful animal, whose ancestry could not be traced to the Mongolian or Arab breeds which it most resembled externally. Internally, the animal is unique in the way its heart and lungs have become enlarged to cope with high altitude. Peissel estimated the nomadic horsemen of this region have been breeding and training the animals for over 1,400 years.

Peissel went back in 1995 to study the breed further. He was unable to bring any examples back with him, but the trip resulted in an unexpected bonus discovery. While traversing the remote northeastern Tibetan region of Riwoche, Peissel found another new horse.

The Riwoche horse bears no resemblance to the Nangchen. Peissel described it as "pony-sized, a little like a donkey but with small ears and a rough coat. It has a black stripe down its back, stripes on its back legs, and a black mane. I thought it looked like cave drawings of horses." Dr. Ignasi Casas, a veterinarian on the team, added, "It looks very primitive and very tough." With their "truncated, triangular heads and very bizarre narrow nostrils and slanted eyes," as Peissel described them, these horses resembled no living breed. He believed they might be the most ancient, unchanged equine race on Earth, a living reminder of the Stone Age.

The Riwoche animals are semi-domesticated, running loose most of the time. The local inhabitants rope them when needed to use them as pack animals or mounts. These horses are described as a relict population, inhabiting a valley almost cut off from travelers. (Peissel and his expedition stumbled on them only because the men were seeking an alternate route home, the pass they intended to use having been blocked by snow.)

In size and general conformation, the Riwoche animal resembles Przewalski's horse, previously Asia's last known wild equine. While Riwoche horses do differ in the shape and appearance of their heads, there are enough similarities to suspect a link between the two types. Przewalski's is also a small horse, less than fifteen hands (sixty inches) high. It is normally brownish or yellow-gray, with a paler nose and underside, and it has the same black stripe and erect black mane as the Riwoche horse.

Beginning in 1992, several herds of Przewalski's horses have been released in Mongolia after decades in which it was generally assumed the wild animal was extinct. Those existing today (over 1,400, most still in captivity) were carefully bred from a total of thirteen horses known to exist after World War II. While the last definite capture of a wild horse was in 1947, occasional reports have suggested a few wild animals hung on along the Chinese-Mongolian border. It's unknown whether the sightings from this desolate region concern true Przewalski's horses or animals of mixed ancestry. The return of the captive-born animals to the wild, an apparent success, may obscure forever the mystery of whether they still survived on their own.

THE NEW GALAPAGOS

The Caribbean island of Navassa is no one's idea of a tropical paradise. Navassa is an uninhabited, pork-chop-shaped island between Jamaica and Haiti. It consists of two square miles of sun-broiled dolomite, protected by thirty-foot cliffs on every side and covered with yard-wide sinkholes and rash-inducing poisonwood trees.

No one has lived on the island, which is a U.S. territory, since World War II. For decades, its only human visitors were Coast Guard personnel who came ashore periodically to maintain a navigation beacon. The beacon was shut down in 1996, and Navassa was left to the small creatures which scurry through its underbrush or hide in the crevices of its rocks.

Save for one expedition in 1930, Navassa had never drawn the attention of biological scientists. That changed in 1998. When a team from the Center for Marine Conservation (CMC) paid an extended visit to conduct a thorough survey, the visitors were left marveling at the diversity of the island's fauna.

Navassa is home to 800 or more species of plants and animals, many endemic to the island. As many as 250 species may be new to science. The CMC expedition, plus a follow-up voyage in April 1999 sponsored by The Discovery Channel, collected an impressive variety of small animals which seemingly had no business existing on this barren rock. One scientist dubbed Navassa "the new Galapagos."

Dives off the coast netted five new fish. These included a new species of blenny, with spines jutting from its head, and a uniquely marked new clingfish only an inch long. On land, there were twelve species of lizards (one scientist estimated the island should only have supported one species), a blind snake, and twenty-five new spiders, one belonging to a new genus. Arachnologist Giraldo Alayon has so far identified two new scorpions to add to the spiders.

Neither expedition found any mammals, but that wasn't unexpected. For a mammal species to be established on the island, a pair (or a pregnant female) would have to first be lucky enough to arrive at the remote island in good health after swimming or drifting. Any mammal reaching land would then have to scale the cliffs. Human visitors come ashore at Lulu Bay, where they must climb a wire-rope ladder fixed into the stone ramparts. A mammal has less chance than a smaller lizard or insect of finding enough suitable fissures and footholds to clamber to safety, and there is no evidence so far of any mammal save man surviving the ordeal.

The abundance of species (and the plethora of new ones) found on Navassa reinforces the importance of searching for new animals in unlikely and inhospitable places. Navassa is a testament to Ian Malcom's mantra in the film *Jurassic Park*: "Life finds a way."

A PARADE OF MONKEYS

South America's largest nation, Brazil, fittingly boasts the world's largest tropical rainforest. Within that forest, the dense foliage and multi-story canopy conceal an amazingly diverse collection of animal species. The great 19th-century naturalist, Alfred Russell Wallace, wrote of this region, "While there appears at first to be so few of the higher forms of life, there is in reality an inexhaustible variety of almost all animals."

Wallace's observation is especially applicable to primates. As of 2004, over ninety Brazilian primates had been identified. Many of those are very new to science. In the last two decades, previously unknown primates have seemingly been falling out of the treetops.

The recent finds started with a new squirrel monkey, a two-pound creature presented in 1985 with the scientific label *Saimiri vanzolinii*. This primate inhabits a tiny territory at the joining of the Amazon and Japura rivers in central Brazil.

The next discovery involved an even smaller creature, now known as the black-faced lion tamarin. This monkey, not much larger than a rat, was found on a coastal island named Superagui. Professors Maria Lucia Lorini and Vanessa Guerra Persson described the species in 1990 as *Leontopithecus caissara*.

The next new primate emerged from the depths of the rain forest. In the fall of 1992, the first report of the Rio Maues marmoset was published. Marmosets are classified with tamarins in the family of small primates called Callitrichidae. There are at least twenty-six species in the family. This group, which ranges from South American rain forests up into Central America, includes more than a quarter of the species in the class of Platyrrhines, or New World monkeys. The Rio Maues marmoset is a long-tailed grayish furball weighing less than a pound. *Callithrix mausei* sports faint black stripes, large eyes, distinct ear tufts, and a cuddly face a bit like a koala's.

Primatologist Russell Mittermeier, president of Conservation International, commented in 1992 that Brazil might yield as many as four or five additional new primates. As events were to demonstrate, Dr. Mittermeier was being too conservative.

Later in the same year, a wildlife survey in that year near the town of Humaita netted a specimen which was named *Callithrix nigriceps*, the black-headed marmoset. This monkey apparently has a very restricted range in an area now subject to heavy development. *C. nigriceps* is part of the subgenus known as the bare-eared marmosets, distinguished by the absence of the hair tufts which many marmosets sport in and around their ears.

A larger primate discovered the same year was *Cebus kaapori*, the fifth known species in the capuchin genus. The Ka'apor capuchin was found in the state of Maranhao. Not surprisingly, it was already known to the local Urubu-Ka'apor and other indigenous tribes. At least one specimen was being kept as a pet at the time Helder Queiroz "discovered" the capuchin on behalf of science. This monkey has a larger area to roam than *Callithrix nigriceps*, but it, too, is under heavy pressure from developers, squatters, and hunters.

In 1996, the Satere marmoset (*Callithrix saterei*) was described. Dr. Jose Ayres of the World Conservation Society first reported the existence of an unclassified marmoset in 1994, and an expedition collected the type specimen in the Brazilian state of Amazonia. Like most of its relatives in the family Callitrichidae, the new find was a small primate with a long tail. Unlike every known marmoset, however, this little black-eared monkey boasted "a distinctive face and ears, unpigmented facial skin, mahogany colored fur, and fleshy appendages on the genitalia of both sexes whose purpose has biologists puzzled," as Les Line wrote in the *New York Times*.

The next monkey species to be formally described from Brazil emerged in 1998. Dutch primatologist Marc van Roosmalen called his find the black-headed sagui dwarf marmoset. It is the second-smallest known primate, with a body about five inches long. Dr. van Roosmalen believes he has several more new primate species collected and awaiting description, along with an impressive array of other mammals (see the next essay in this section.)

The dwarf marmoset was named *Callithrix humilis*. This description was originally based on a single specimen that a Brazilian whom van Roosmalen had never met brought to his doorstep one morning. The van Roosmalens' home has become a sanctuary for orphaned monkeys, and Marc and his wife took in this one as well. When van Roosmalen looked into the milk can holding the diminutive primate, he was surprised to see how unfamiliar it was. The animal was grayish with a black tail, black fur on top of its head, and a fringe of white surrounding its face. Van Roosmalen was excited but tried to keep calm. He was afraid that if his visitor knew the monkey was important, "the guy maybe wouldn't have given it to me."

The creature's uniqueness was confirmed several weeks later, when Russell Mittermeier visited his colleague. Dr. Mittermeier took one look at the animal and proclaimed, "It's new!" Van Roosmalen and others quickly began to search for more marmosets like the one they called "Dreumes" ("Little Fellow" in Dutch), but it took more than a year before any were found.

Unlike other marmosets, the new species is not territorial. In the known marmosets, only a dominant female in the troop gives birth: another rule apparently cast aside, or not yet developed, in the latest type. Van Roosmalen told an interviewer, "I think the DNA will show it's at the start of the (marmoset) radiation, a survivor relic. It may be the missing link to understanding that radiation."

Finally, taxonomic revisions in the last decade have produced the claim that two other species, *Callithrix intermedia* and *Callicebus hoffmannsi*, were collected some time ago but not recognized as distinct types. It is clear we have much to learn about our little cousins, the monkeys.

VAN ROOSMALEN'S MAMMALS

Dr. Marc G. M. van Roosmalen, the primatologist we met in the last essay, is a very busy and successful discoverer of Brazilian mammals—and not just primates.

He certainly has found plenty of primates, beginning with the dwarf marmoset just mentioned. While van Roosmalen originally described this animal as a new species and placed it in the genus Callithrix, he has since come to believe it is even more distinctive. It neither looks nor acts like other members of Callithrix. The new species has single offspring, often has more than one reproductive female in a group, and is non-territorial—all exceptions to the Callithrix rules. Van Roosmalen has since published a paper naming a new genus, Callibella.

This find only whetted Van Roosmalen's appetite for understanding the primate fauna of his adopted country. In March 2000, he reported, "I [have] collected data and brought home live specimens of another thirteen new primates, most full species." Shortly thereafter, he published descriptions of the latest two, labeling them *Callithrix manicorensis* and *Callithrix acariensis*. The first, also called the manicore marmoset, is about the size of a squirrel. It's a colorful creature, with a white or silvery body set off by a black tail, and a yellow or orange underside. The second, the Acari marmoset, lives in a small territory adjoining that of its genus-mate. It's almost as uniquely colored, with a mainly white body, a gray back, and a black tail with an orange tip.

Another new monkey Van Roosmalen has collected, announced in 2002, he christened *Callicebus benrhardi*. Reddish-orange with a brown back and a dot of white on the end of its tail, the Bernardi monkey is about sixteen inches long and weighs just over two pounds. Local people have given it the delightful name of "zog-zog." This find was followed by *Callicebus stephennashi*, or Stephen Nash's titi monkey. The new titi is another colorful character, with a silver coat, a black forehead, and red underside and sideburns.

Van Roosmalen has collected examples of what he believes are one additional new species in Callicebus and two new species of uakaris. One of these is white, the other a dark brownish-black. He has also either published descriptions or is preparing them for several more types, including a new orange woolly monkey (named by him *Lagothrix jutaiensis*) and two new species of spider monkeys, the largest known South American primates.

It may be that not all the monkeys van Roosmalen believes are new species will turn out to be so distinct. Variations within a species can be considerable (just look at humans). Still, the diversity of his finds is striking. All this comes, he has noted, "in an area thought not to be very important."

Leaving the monkeys, van Roosmalen's long studies in Brazil have also netted him a new species of dwarf porcupine. The porcupine has a black tail, a pink nose, and soft-looking pale hair covering its spines. It has been christened *Coendu roosmalenorum*. The type specimen was being kept as a pet by a girl in a tiny settlement called Novo Jerusalem on the Rio Madeira. To accompany the porcupine, so to speak, there is van Roosmalen's new five-banded armadillo.

The busy primatologist has also found evidence for a new peccary, or wild pig. Van Roosmalen has seen this animal twice. The peccary lives in small family groups and sometimes joins herds of the two known species of this region, the collared peccary (*Tayassu tajacu*) and the white-lipped peccary (*Tayassu pecari*). Van Roosmalen's peccary appears to be intermediate in size between the other two species, which puts the adult weight in the range of fifty to seventy pounds. Van Roosmalen describes the new animal as "dusty brown." The white-lipped peccary may be brownish, but has distinctive white markings. The collared peccary is uniformly black-black. In 2004, an IUCN website carried a photograph of the animal "thought to be the fourth species of peccary".

Van Roosmalen also believes there is an unclassified deer in the rain forests. The most spectacular reports he's heard, though, concern what may be a new species of jaguar. Hunters have described the cat to van Roosmalen as solid black in color and slightly larger than the known jaguars of this region. It reportedly hunts in pairs. While black (melanistic) jaguars are not uncommon, van Roosmalen's informants claim this is a different animal altogether, one distinguished by a white throat and a tuft on its tail like a lion's. There have been occasional reports of such a "black tiger" from Brazil going back to the 1700s, when naturalist Thomas Pennant illustrated it as a cat with solid black body (that is, with no partly obscured black rosettes, as are seen in the melanistic jaguar) and a much lighter underside. If confirmed as a species, this would be the first new big cat named since 1858. Even if it's only a new color morph of a known species, it is certainly an interesting one.

Dr. van Roosmalen's labors are evidence of what one determined man can contribute to science—and how many mammals are still waiting for equally dedicated researchers to seek them

Roosmalen's black jaguar

out. Along the way, van Roosmalen has won many victories for conservation and is rediscovering Stone Age farming methods that may reduce the drive to slash and burn the rain forest. His selection in February 2000 as one of *TIME* magazine's "Heroes for the Planet" is a well-deserved honor.

MARINE LIFE BY THE NUMBERS

There are 317,000,000 cubic miles of water in the world's oceans, and only a tiny fraction of this stupendous volume can be monitored by the instruments of science. This water is spread an average of two miles deep over an area of 143,000,000 square miles. Given the contours of undersea mountains, canyons, and other features, the actual area of the seafloor is much greater.

How much life exists in the ocean depths? Nearly three thousand years ago, a Psalmist described the sea as a place "wherein are things creeping innumerable, both small and great beasts." He was right.

It was once estimated there were only 200,000 species of marine life of all kinds. By 1995, William J. Broad, science writer for the *New York Times*, reported the estimates for life on the sea floor had risen to ten million species or more. Said one marine biologist, "It's changing our whole view about biodiversity."

Most of these creatures are tiny, like the species in the new phyla Cycliophora and Loricifera mentioned earlier. Many fall into a category called meiofauna: animals which live on the seafloor and measure less than one millimeter (1/25 of an inch). In one experiment, a thorough sampling of an area of the seafloor "the size of a large living room" at a depth of 13,000 feet produced 798 species, 460 of them new.

Some new animals are larger than those in the meiofauna class. For example, a new species of giant sea louse a foot long was recently collected off Australia. To visualize this creature, imagine a common wood louse blown up to the size of a man's shoe.

In 1998, scientists at the Monterey Bay Aquarium Research Institute (MBARI) discovered a startling presence over 2,000 feet down in the Monterey Submarine Canyon. A jellyfish, three feet across and colored a striking blood red, was floating through the depths. "Big Red" had none of the usual jellyfish tentacles, only a variable number (four to seven) of heavier "oral arms." *Tiburonia ganjoro*, formally described in 2003 and requiring a new sub-family, remains a mystery in many ways. "We know almost nothing about it. What it does. What it eats. What eats it," said MBARI's George Matsumoto.

Jellyfish (or sea jellies, as marine biologists prefer to call them) are a ubiquitous group that we still have much to learn about. They drift or swim through all the oceans of the world, and may be found at the surface or thousands of feet below. Some glow softly: others sparkle like miniature chandeliers. They may be shaped like hemispheres, spheres, cones, or countless variations of these. Jellies can be deadly, like the sea wasp (*Chironex fleckeri*), or huge, like the Arctic lion's mane (*Cyanea capillata*), known to be up to seven feet across with 120-foot tentacles. Jellies play an important role in the marine food chain, and new species are found every year.

Big Red's discovery came only a year after the formal description of *Chrysaora achlyos*, also from the eastern Pacific. That *Chrysaora achlyos* remained unknown so long is inexplicable, as it's a surface dweller that has been seen in large numbers. Not only does it have a purple bell three feet across, but it sports pleated oral arms extending twenty feet and tentacles trailing well beyond that. A very different jelly was described in 2004. *Stellamedusa ventana*, collected by MBARI scientists from depths of 500 to 1,800 feet off California, was weird enough to merit creation of its own subfamily. It is a beautiful creature, about four inches across with a translucent, blue-white body shaped like a bell flattened at the top. It glides through the depths like the ghost of a comet. Tiny bumps (actually clusters of stinging cells) cover the oral arms and the surface of its body.

There are so many unknown species in the sea that it's easy to find them without looking for them. In the course of making a documentary series, *The Blue Planet*, released in 2002, videographers captured images of two bizarre new animals from the depths. These were the hairy angler, a fish the seize of a beach ball and studded with a forest of antenna-like projections for detecting the movements of prey, and an octopus they christened "Dumbo" for the two large, round flaps projecting from its large, round head.

Seamounts, extinct underwater volcanoes, are another source of diversity. One survey of twenty-five seamounts in the Coral and Tasman Seas netted 850 species, at least a third of them new to science. The authors of this study, published in June 2000 in *Nature*, noted that there are an estimated 30,000 seamounts in the world, the vast majority of which have never been examined by scientists. There is no doubt examining new seamounts will unveil new species, since these features act as undersea "islands" on which animals often develop in isolation. Tony Koslow, one of the seamount researchers, reported that, "We found not a single species in common between the seamounts off Tasmania and those

3,000 kilometers (1,875 miles) to the north, which is extremely unusual for the deep sea, where most species have extremely wide distributions, typically over much of an ocean basin."

It's no exaggeration to say that every expedition probing the ocean depths finds new species. A far-ranging sampling of the Mid-Atlantic Ridge in 2004 by scientists from the Institute of Marine Research in Bergen, Norway, reported new species of fish and invertebrates, strange burrows on the seafloor made by a still-unknown species, and a bizarre creature about a foot long which looked gelatinous but had a well-defined head (or forepaw) and tail. The expedition was unable to bring up the thing seen crawling across the bottom 6,500 feet down, and they have no idea how to classify it. It may represent a new order or class of animal.

Many fish, large and small, are among the animals awaiting discovery. Dr. Dominique Didier Dagit, an ichthyologist with The Academy of Natural Sciences in Philadelphia, identified a striking new species from the South Pacific in 1998. Dagit is the world's pre-eminent specialist on the chimaeroid fishes. Chimaeras, large cartilaginous fish with oversized heads, large pectoral fins, and long, thin tails, are also known as ratfishes. They are deep-dwelling species (found from 1,000 feet to over a mile down) about which very little is known. Dagit explains that, "The reason I study them is because nobody else does. There are so many species that are undescribed. We don't know about their distribution or their reproduction."

Dagit was visiting the National Museum of New Zealand when the staff asked her to take a look at an odd specimen brought in by a fisherman. It turned out to be a new species, which she described as the leopard chimera, *Chimaera panthera*. The twelve-pound fish is over three feet long and is covered with distinctive brown spots and blotches. In 2002, Dagit named two new species: *Hydrolagus bemisi* and *Chimaera lignaria*. (The latter name means "carpenter chimaera." Dagit named the fish, a large chimaera colored in attractive hues of blue, for her husband, a carpenter.) Dagit believes she has type specimens of two more new species in hand to describe, and expects there are several others lurking in the depths.

After the discovery of the leopard chimaera, Dagit commented, "This discovery indicates there are more things out there than we ever knew. We have only discovered a fraction of the species living in the oceans."

A massive effort called the Census of Marine Life is underway to determine more precisely what that fraction is. The ten-year project, funded by government and private donors to the tune of a billion dollars, has catalogued 15,304 marine fish and expects to add as many as 3,000 more by 2010. Already, scientists funded by the project have collected over 600 new species. Hundreds, if not thousands, of additional new species are expected.

Marine life exists far deeper than scientists once believed. In the 1990s, new underwater craft built in Japan repeatedly probed new depths—and found new life.

The submersible *Shinkai 6500* found a colony of clams at a record depth of 20,000 feet, which was remarkable enough. Scientists on the same submersible at about the same depth in 1997 videotaped a new species of five-inch-long transparent worm, apparently belonging to a group called the polychaetes. (The *Shinkai 6500* holds the depth record for an untethered craft carrying humans: 21,320 feet.)

In 1995, the robotic submersible *Kaiko* settled to the bottom of the Challenger Deep, 35,798 feet below the surface. In a half-hour visit, the craft's lights picked out a worm, a shrimp, and a sea slug. This confirmed the findings of life at these ultimate depths made by the bathyscaph *Trieste*, which descended into the same area in January 1960. The Trieste's crew, Don Walsh and Jacques Piccard, reported a foot-long flatfish moving across the bottom. Most authorities believe this was a misidentified holothurian (sea cucumber), but it was still life.

The deepest a fish has been recovered from the seas is 27,453 feet. This was a new species, *Abyssobrotula galatheae*, caught in 1970 and described in 1977. The six-inch fish had been scooped from the Puerto Rico Trench in the Atlantic.

While animals of every size are important to science, it's the truly large creatures which capture the public's imagination. In 1998, using a statistical technique employed by biologists to estimate the diversity of animal populations, Dr. Charles Paxton of Oxford University tried to calculate the number of animals measuring two meters (six and a half feet) or longer likely to exist still unknown in the world's oceans. The resulting estimate was forty-seven species. Paxton did caution that estimating the number of unclassified animals by studying the discovery rate and total number of such species found so far is very inexact. He guessed the oceanic types still to be found include whales and sharks, although it is possible that some might be totally new types of animals. Whether this proves true or not, Paxton has given fellow scientists a good reason to keep scanning the waters.

Marine biologist and explorer Sylvia Earle summed up the situation very well: "The reality is we know more about Mars than we do about the oceans."

THE AMAZING OCTOPUS

Nothing found under the seas is stranger than a recently discovered Indo-Pacific octopus. In the shallow waters of three straits off Sulawesi, Indonesia, lives an animal with an armspread of about two feet and an astonishing talent. The Mimic Octopus, which has long, slender tentacles and is normally dark brown with white stripes and blotches, has powers of imitation unparalleled in the natural world.

The octopus can curl up its tentacles at its sides and darken itself to resemble a stingray, with one tentacle trailing at the back to make a tail. It can turn tentacles into fake pectoral fins and look like the head and forebody of a jawfish rising from the seafloor. It can arrange its entire shape and color to mimic a flounder and glide across the bottom. It can resemble a starfish, a jellyfish, a sand anemone, a sea snake, a snake eel, a lionfish, or a baby cuttlefish.

Where does this little octopus learn such complex behavior, and what is this repertoire used for? The octopus can imitate both prey species and predators, so it may use this ability as a defensive mechanism or a way to sneak up on prey, as the need arises. However this behavior arose, and however it is passed on, it is a most impressive example of skill and adaptability among these "lower" animals.

Readers who (understandably) believe I am exaggerating this cephalopod's abilities are directed to Roger Steene's 1998 book *Coral Seas*, where the full range of the animal's talent is displayed in beautiful color photographs. The same book, incidentally, will introduce the reader to other recently described species including: a stunning blue porcelain crab found inside a sponge; a neon-brilliant nudibranch, an inch and a half long, from northern Sulawesi; and a completely transparent shrimp discovered by accident while Steene was focusing his lens on the bubble coral on which the shrimp was resting.

Dr. Mark Norman of James Cook University in Queensland believes he and his colleagues have nabbed no fewer than 150 new species of octopus in the last ten years. Incidentally, Norman recently explained why the Mimic Octopus has not been formally described, even though it's not very hard to find: "Everybody likes them too much to knock one off for formal description." He is preparing a description of another long-armed species he calls "Wonderpus," an Indonesian native which can mimic sea snakes. (The Wonderpus and Mimic Octopus are occasionally confused in media reports, but they are two separate types.) Also in work is a

Dr. Mark Norman's mimic octopus, and its Indonesian relative, the "Wonderpus."

description of a species with very thin limbs three feet long, known to Norman as "Spaghettiopus."

In the meantime, the mimic is already in trouble, as the first specimens of this undoubtedly-rare type have begun turning up in the largely unregulated international aquarium trade. There is currently no national or international legal protection for any form of octopus, a situation that frightens some marine biologists.

One new species of octopus did get a lot of attention in the media—not for its scientific importance, but for its behavior. In December 1993, fortunate scientists aboard the submersible *Alvin* videotaped a mating encounter between octopuses on the seafloor over a mile and a half below the surface of the eastern Pacific. The creatures were of different species, which made this scene unusual enough, but the press attention erupted because the octopuses in question were both males. Apparently, encounters between cephalopods at this depth are so rare that an animal will attempt mating just on the chance that what it's bumped into is a receptive member of the right species and the opposite sex. (This is not what we humans would call "safe sex," since this strategy also entails some chance of the smaller participant's being killed and eaten if it's guessed wrong about the identity of its prospective partner.)

In any event, when the sensationalism over "gay octopuses"

died down, Dr. Janet Voight of the Field Museum of Natural History found time to actually study the octopuses she had helped capture on tape. The larger octopus, a grayish animal, had slithered off in the darkness and has not been seen since. The smaller one (about fifteen inches across), while unknown at the time of the video, did turn up in the form of several specimens and additional videotapes. Incredibly, it seems to favor the waters around hydrothermal vents—waters containing lethal concentrations of hydrogen sulfide. How the octopus protects itself from this poison is still being studied. The species was described in 1998 as *Vulcanoctopus hydrothermalis*.

The description of *V. hydrothermalis* created a new genus as well as a new species. This was understandable, since the animal has several distinctive properties in addition to its habitat. First, its skin is transparent. In fact, the animal is almost lacking in pigment. Except for its black eyes and purple hearts, all the creature's tissues are either see-through or white. The octopus even has a strange build, with a tiny head and exceptionally long, slender arms.

In 2005, a major breakthrough in cephalopod research was announced. Japanese scientists lowered a baited camera trap 3,000 feet down in the North Pacific Ocean to capture the first images ever taken of a live giant squid of the genus Architeuthis. The number of cephalopods we have yet to locate and identify is anyone's guess.

Countless other invertebrates have been found on the seafloor, with more constantly being described. One relatively recent find is a foot-wide oyster with a massive shell from Palau. Its discoverers refer to the archaic creature by the overused but always fascinating term "living fossil." Its line was thought to have vanished millions of years ago, but live specimens were found in the 1980s on reefs off Okinawa.

A new soft coral (that is, a coral without a calcified skeleton) from the same region "inflates" with water to extend white flowerlike structures into the current, but retracts into a little ball when water is expelled. Looking even more like a flower is another coral, *Javania exserta*, described in 1999. *J. exserta* is an ahermatypic coral, a kind which lives individually and does not build structures. This particular species expands into a delicate pinkish-red bloom which spreads out a couple of dozen tentacles. The tentacles are actually transparent, but each has a narrow white stripe running down its length and a dot of orange at the tip.

In 1997, Patrick Colin collected a very unusual deep-sea starfish, also from Palau. (As with jellyfish, biologists are trying to discourage the inaccurate name "starfish:" "sea star" is preferred.) Nicknamed the "cornbread star," the animal is orange and white and so thick-bodied that it looks like an inflatable toy or a stuffed pillow. It is over a foot across and exists only in the zone from 250 to 420 feet.

NEW BIRD IN NEW ZEALAND

The island nation of New Zealand was, thousands of years ago, a paradise for birds. With not a single mammalian predator anywhere, the birds assumed a dizzying array of forms, colors, and ecological niches.

When the Maori people arrived perhaps a thousand years ago, they and their accompanying dogs and rats took a heavy toll on the birdlife. Eventually Europeans, with their guns, farming methods, and pet cats, appeared to make things much worse. Many species, like the huge, flightless moas, were driven into extinction. Modern New Zealand wildlife officials have tried to keep an accurate inventory of the surviving species and transplant as many as possible to mammal-free offshore islands. It is this effort that brings us to our tale.

Amid the carnage of the island's birds, there were few species left alive and undiscovered by the 20th century. It was 1930 when the last new bird species was described. That, ornithologists widely assumed, was the end of such discoveries. The only bright spot for a long time was the 1948 rediscovery (or re-re-rediscovery) of the takahe, a colorful, turkey-sized ground dweller that was declared extinct three times but always found some remote spot in which to hang on.

In 1997, a Department of Conservation team traveled to Jaquemart Island looking for a rare species, the Campbell Island teal. Instead, one of their bird-sniffing dogs flushed out an unknown species of snipe. About ten specimens were located during the expedition. The first new bird in almost 70 years was a heartening discovery for ornithologists used to watching species disappear. Finds like these remind us that a long time without discoveries, even in a damaged ecosystem, does not mean there are no discoveries left to make.

Another Pacific island that does not see many modern discoveries is the populous nation of Taiwan. A medium-sized bush-warbler with a slender bill and rather drab brown coloration had been known from the island since at least 1917. Ornithologists assigned it a variety of identities, but always as a local population of an Asian mainland species. Four ornithologists led by Pamela

Rasmussen of the Smithsonian Institution recently took a closer look and discovered the mountain-dwelling Taiwanese bird was actually unique. Accordingly, the description of the Taiwan bush-warbler, *Bradypterus alishanensis*, was published in April 2000. It is Taiwan's fifteenth known endemic bird species.

This was only one of Rasmussen's recent descriptions of new species. In 1999, she published a paper naming the cinnabar hawk-owl, *Ninox* ios, from Sulawesi. This discovery came when Rasmussen spotted a misclassified specimen in a museum in Leiden, the Netherlands, and realized it represented an unknown type. In 1998, she co-authored the description of the Sangihe scops-owl, *Otus collari*, and named the Nicobar scops-owl, *Otus alius*.

From the island of Mindanao in the Philippines came another discovery. In 1997, three ornithologists described *Aethopyga linaraborae*, which has the charming common name of Lina's sunbird. The bird was identified from specimens taken on three mountains on the eastern edge of the island. As so often happens, the first specimens had been in a museum a long time—since 1965—and had never been properly identified. Lina's sunbird is a colorful creature, with a green back, yellow breast, and emerald green and ultramarine wing feathers. Of Mindanao's seventeen endemic bird species, no fewer than sixteen were 20th-century discoveries. Further north in the same island group, a rare and apparently flightless new species of rail was spotted by a biological survey team on the island of Calaya in 2004. The Calayan rail, *Gallirallus calayanensis*, has a population estimated at 200.

The recent finds put an exclamation point on a 1998 analysis by ornithologist A. Townsend Peterson. Peterson noted how a leading zoologist, Ernst Mayr, had predicted in 1946 that fewer than 100 bird species remained to be found in the entire world. In the period 1941 to 1997, however, a minimum of 163 valid new species had been added, and the rate of discovery actually seemed to be increasing. (For the record, Mayr admitted in 1971 that he'd underestimated avian diversity. It was one case where a scientist was happy to be wrong.)

Peterson credited much of the recent success in finding new species to the late Ted Parker, who had pioneered in analyzing birdsongs as a way of distinguishing between species and detecting new species. Parker, who also co-developed the concept for CI's RAP teams, could recognize thousands of species from song alone. This talent led to the discoveries of several new birds.

Parker died in a plane crash in 1993 while pursuing his work, but others are carrying it on. There is, Peterson noted, "no sign of exhausting the supply" of new species.

MONITORING NEW LIZARDS

The monitors are the largest lizards in the world. They include the reigning lizard heavyweight champion, the Komodo dragon, which may be ten feet long and has been known to include humans in its diet. The monitors are a highly successful group distinguished from other reptiles by their active hunting style (often compared to that of mammalian predators) as well as their size.

Interested researchers like German zoologist Wolfgang Boehme have only recently documented just how widespread and diverse monitors are. A total of sixteen species have been described from Indonesia and surrounding islands since 1990. Australia's count has also risen recently, up to twenty-seven species, with others known but yet to be named.

Smaller monitors are popular as pets, a fact which has led to several recent discoveries. A 1998 article by Jeff Lemm in *Reptiles* magazine discussed eight new types, all of which have turned up in the pet trade. Two species have been formally described. There may be as many as six more species awaiting description. Lemm cautioned that some of these may prove to be subspecies or varieties rather than species.

From the island of Halmahera came *Varanus yowoni*, the black-backed mangrove monitor, which may be five feet long. The other described species Lemm mentioned was the smaller, bright-yellow quince monitor, *Varanus melinus*. This lizard grows to almost four feet in length and is reportedly a very tractable species in captivity. It was kept as a pet and even bred in the United States for years before being identified as a species.

In 1999, Dr. Robert Sprackland described another species, the peacock monitor *Varanus auffenbergi*. This is a colorful animal, blue with turquoise and orange markings, plus bright yellow spots on its limbs. Smaller than the other new monitors, it was obtained from Roti island in Indonesia.

Some new species are much harder to place. It's often the case that a type specimen is in hand, and yet the type location is unknown. Because of the mixing of sources and suppliers in the pet business, it remains uncertain where several recently described species of new monitor lizards originated. Wherever they came from, the most interesting thing is that, until recently, the scientific world had been quite unaware of them.

A TURTLE—OR THREE

A new turtle from Queensland, Australia, has astonished herpetologists who thought its kind went extinct over 20,000 years ago. The zoological world's newest "living fossil," the Gulf snapping turtle, was rediscovered in 1996 when someone brought a turtle shell to herpetologist Scott Thompson and asked him to identify it. Thompson compared it to a fossil known from the same locale and found they had the same unique configuration of plates on the lower shell, or plastron. The shell of the rediscovered turtle is about thirteen inches long. The fossil was three inches longer but otherwise identical.

Thompson told a reporter he expected further discoveries from northern Queensland. "It's in the middle of bloody nowhere and so we humans ignore it. Lots and lots of new species are going to come out, not just turtles."

The discovery of a related species is an even stranger story. The tale of a new tortoise began in 1963, when an Australian scientist, John Cann, spotted hatchlings in Sydney pet stores which belonged to an unknown species. Despite getting no help from the store owners, Cann eventually tracked down a supplier in Queensland who was collecting the eggs from the wild. It took Cann until 1990 to actually locate an adult specimen. The Mary River tortoise, with a fifteen-inch-long shell and a long, heavy tail, proved to be deserving of a new genus. Cann and an American colleague, John Legler, named the genus Elusor in honor of the animal's success in hiding from the view of science.

More recently, the Burnett River snapping turtle was described, also from Queensland. Australia's newest chelonian is its heaviest snapping turtle, weighing over fourteen pounds. It is also known for the strange ability to absorb oxygen through its cloaca, meaning the animal could theoretically bury its head in the mud and keep it there as long as the hind end was above water.

A fourth recent turtle discovery, from Southeast Asia, is in many ways the most surprising of all. Hoan Kiem Lake is a tiny, polluted body of water only seven feet deep in a park in the Vietnamese capitol of Hanoi. Despite the lake's size, it has long been associated with legends of giant turtles. A large, dried-out turtle, found dead about thirty years ago, is maintained in a temple on an island in the lake. Still, most scientists paid no heed to the legend until a video cameraman taped a turtle emerging from the water in 1998.

Ha Dinh Duc, a biology professor at Hanoi's University of Science, had long believed there really were outsized turtles still living in the lake. After studying the video, he announced the Hoan Kiem turtle was a unique species, heretofore undescribed and almost unknown despite its habitat. Duc estimated the turtles could be six feet long and weigh over 400 pounds. This would make them far larger than any known freshwater turtle. Most references give the title of "largest freshwater turtle" title to the giant river turtle of the Amazon (*Podocnemis expansa*), which weighs in at a comparatively puny 150 pounds. A five-foot-long Hoan Kiem turtle estimated at 550 pounds was photographed at close range in March 2000.

Some turtle experts think Duc is overreaching the evidence, and that any turtles in this lake are probably members (perhaps unusually large ones or a new subspecies) of a known species. The candidates put forth are the Asian giant softshell turtle (*Pelochelys bibroni*) and Swinhoe's softshell turtle (*Rafetus swinhoei*), both of which may rival *P. expansa* in size.

Dr. Peter Pritchard, one of the world's leading turtle experts, doesn't agree the Hoan Kiem turtles can be assigned to a known species. According to Pritchard, photographs indicate the shape of this turtle's head and of its carapace are unique. He agrees with Duc, who has named the turtle *Rafetus leloii*. The genus Rafetus, also known as the Bicallosite Softshell Turtles, includes several species, all from Asia. Duc has based his species on the stuffed turtle in the Ngoc Son temple and a skeleton in a Hanoi museum.

If the Hoan Kiem is a species, Pritchard fears, it may be a doomed one. There are no more than four or five turtles in the lake, and there is no evidence they are reproducing. He believes if the lake were cleaned up and a sandy beach for egg-laying added, the animals still might recover. Some researchers are even more despairing, suggesting there might be only one turtle now living.

Anders Rhodin co-chairs the Tortoise and Freshwater Turtle Specialist Group of the World Conservation Union. He is among those who think the Hoan Kiem animal is something unique and valuable. "This species is a huge, huge animal that's incredibly endangered and it really needs help," he said. Vietnamese and American researchers plan to visit other lakes where giant turtles have been reported, hoping the turtle survives elsewhere.

The most satisfying resolution to this mystery would come if Pritchard and Duc were proven correct in thinking this massive chelonian is a new species—and if it could be saved. Discovering the world's largest freshwater turtle in the in the heart of a major city would make for one of the most surprising stories in all the history of zoology.

THE VU QUANG FISH

Mammals are not the only new species to come out of Vu Quang and its surrounding region in the last decade. An area where large mammals remain unknown is guaranteed to harbor smaller creatures which are equally new to science. After a Royal Ontario Museum expedition in 1995 brought back a hoard of new snakes, spiders, and insects, Canadian herpetologist Bob Murphy reported, "Vietnam is such an incredible, uncharted country. Vietnam is a huge mystery to us; it's a scientist's dream come true." Murphy returned to Vietnam in 1998 and collected new species of frogs and salamanders, along with more new snakes and a caecilian, a legless amphibian.

Dr. Nguyen Thai Tu, a Vietnamese ichthyologist, agrees. He reported in 1996 that Vu Quang harbors a previously unknown fish of the genus Crossochelius. The new species is eight to ten inches long with a silver underside and a gold stripe down its back. It weighs about three pounds.

Tu has been adding to his country's known fish fauna since 1983. He named his first new fish, *Cobilis yeni*, in that year, then followed up with two species of carp in 1987. In 1992, he found another carp, which he christened *Parazacco vuquangensis*. Accordingly, the most recently described species, known to local villagers as Co and regarded as a common food fish, is Tu's second find from Vu Quang.

The announcement of this discovery included a comment from the WWF's Vietnam Program Director, David Hulse: "Vu Quang is fast becoming a treasure trove of new species. It's vital that we conserve this spectacular region, because there's no telling how many other species we will find."

The modern spate of discoveries in Vietnam may have begun in Vu Quang, but they are continuing in other regions as well. An October 2004 report from the Vietnam News Agency announced an international survey of Phong Nha-Ke Bang National Park in central Quang Binh province had netted over thirty new species of reptiles and amphibians and, at first estimate, a hundred new freshwater fish.

Neighboring Cambodia offers similar opportunities for scientific study. In September 2000, a British-led expedition returned from the Cardamom mountains in southwestern Cambodia. Scientists reported the region was teeming with Siamese crocodiles (*Crocodylus siamensis*), a species thought to be extinct in the wild. According to biologist Jenny Daltrey, the expedition also collected "large numbers of frogs, moths, and subspecies of birds which are almost certainly new. More than a third of the invertebrates we found were new to science." The visitors also heard reports of the mysterious kiting vor, a species described from Vietnam based on its horns (see the previous essay on Vu Quang), but failed to spot what must be the world's most elusive large mammal.

Dr. Daltrey summed the situation up in words very similar to those David Hulse used to describe Vu Quang. She said, "We are going to find many more new species. This is a region of global importance for wildlife."

INSECTS OF THE PAST AND PRESENT

New insects usually don't make news, but it's not every day you find a primitive wasp which has supposedly been extinct for millions of years.

Russian entomologist Alexander Rasnitsyn was browsing the collection at the California Academy of Sciences in San Francisco when an unusual specimen caught his eye. Rasnitsyn noticed peculiar features, including serrated teeth on the ovipositor, or egg-laying tube, on two wasp specimens. To him, this meant they belonged to the same undescribed species as another wasp he'd seen: a fossil, 20 million years old, in a German museum.

The two modern specimens were both from California, one dating to 1937 and the other to 1966. Donald Burdick, the entomologist who captured this second wasp in the Sequoia National Forest, recalled that he knew his catch was something unique, but had no idea how significant it really was. The full scientific description was done by Rasnitsyn and the Academy's entomology curator, Wojciech Pulawski, in 1995.

One of the interesting things about looking for insects is that there are still new species to be found everywhere, including heavily populated areas. In 2001, scientists in the United Kingdom identified that nation's first new butterfly species in 112 years.

The possibility of new insect species in well-known locales makes this a rich field of discovery for new and amateur scientists as well as the experts. In 1983 and 1993, new moths were discovered in largely developed areas of California by University of California Berkeley graduate students. Insect biologist Jerry Powell commented, "It might surprise the public of the Bay Area to know there are any new species still being discovered nearby. You're not going to find any birds, mammals, snakes, but nothing could be further from the truth for insects." Powell estimated a third of insect species in California are still awaiting classification.

A spectacular one turned up in 2002: a new species of Jerusalem cricket resembling an inflated ant, but a full three inches long. Zoologist Robert Fisher marveled, "This is the largest insect by mass in Southern California, and it was undescribed scientifically."

At the other end of the United States, a massive one-day collection effort in the Great Smoky Mountains National Park, on the border between Tennessee and North Carolina, was recently staged as part of a biodiversity project which began in 1998. A small army of lepidopterists collected twenty-five new species of moths in one full day of intensive trapping. Participant David Wagner, of the University of Connecticut, explained the point of this effort: "The first part of conservation is finding out what is there. You can't protect what you don't know about."

One thing no one knew about until recently was an entire order of insects. In 2002, the new order Mantophasmatodea was formally established. These insects were given the fitting common name of gladiators. Gladiators are inch-long wingless insects from Africa, heavily armed with spiked front legs and powerful mandibles (hence the name). Fossils in ancient amber and a never-classified old specimen from Tanzania rested unidentified until a young German entomologist named Oliver Zompro looked at both pieces of evidence and realized just how extraordinary a find they represented. A safari by helicopter into the remote Namib Desert netted the first live specimens of Mantophasmatodea.

As so often happens in science, the gladiators were known long before they were recognized. Specimens collected in South Africa in 1890 were examined by Louis Perinquey of the South African Museum. Perinquey realized that he had a new species, but he pursued the matter no further. When the new order was named based on the Namibian specimens, someone went through the collections in South Africa and brought Perinquey's creatures to light. The long-deceased taxonomist never knew how close he was to a major discovery.

Finally, a discovery from 2003 illustrates that larger terrestrial invertebrates are also still to be found—and in the oddest places. What may be the world's largest leech was collected from a back yard in Salem, New Jersey, in July 2003. "Piwi," as the type specimen was named, is close to 17 inches long when stretched out. A second specimen of the same species was collected and put in Piwi's tank, but that was a bad idea, as Piwi promptly ate it. Dan Shain of Rutgers University is seeking more specimens to confirm and describe the new leech. He may not have to wait long. It seems reasonable that anyone finding a giant black leech on his or her property would be only too happy to have Professor Shain come and haul it away.

ALL THE SPECIES OF THE WORLD

How many species of all kinds are there?

Starting with the vertebrates, lists compiled in the 1990s of known species included 4,675 mammals, 9,702 birds, 7,870 reptiles, 4,780 amphibians, and 23,250 fish. In 2000, the annual rate of discovery for the latter three groups was estimated at sixty reptiles, eighty amphibians, and 200 fish. Rates for mammals, as we have seen, range as high as forty-five per year if reclassifications are counted. New birds are the rarest, but, even so, new species have been added at rates of two to four per year since 1980.

When we get to the invertebrates, the numbers get out of hand. The number of described invertebrate species is well over the one million mark, but the count changes so quickly that it's more illustrative to cite examples than to offer a total which will be obsolete long before the reader sees this book.

Taxonomists describe some 13,000 new species a year, the vast majority being arthropods or other invertebrates. About 6,000 of these are insects. Entomologist Terry Erwin once found 1,100 species of beetles on nineteen trees of one species in Panama. From those, he extrapolated there may be thirty million species of tropical beetles, although some of Erwin's colleagues consider such a figure highly speculative.

An insect survey in Sulawesi, Indonesia, netted 1,690 species, of which almost two thirds were undescribed. In 1994, just one organization (The National Biodiversity Institute) in one country (Costa Rica) reported it was discovering 300 new species per month in the course of its search for medically useful compounds.

This last example was provided by Dr. Dan Janzen of the University of Pennsylvania in 1994. Janzen, a biology professor specializing in tropical biodiversity, also offered some interesting estimates of the number of animals yet to be discovered: 50,000 vertebrates, 150,000 crustaceans, and 200,000 mollusks (according to another source, there are 600 species of mollusks discovered or rediscovered every year). Also believed still out there are 750,000 to 1,000,000 spiders and mites and anywhere from 8,000,000 to 100,000,000 insects. The United Nations' Global Biodiversity Assessment in 1995 offered a "working estimate" of 13.6 million animal species of all types. (The same report noted that 5,400 species are known to be under threat of extinction. Since most species are unknown to begin with, the true number must be much higher. How much higher is a subject of considerable controversy.)

All these numbers are only educated guesses, but no one doubts there are millions of insects and other small animals left to identify. An enormous undertaking, the All Species Inventory, was begun in 2000 by the private All Species Foundation. The goal of this audacious project is to categorize every living thing on the planet within twenty-five years. This is not only a challenge of logistics, but of brainpower. While new tropical beetles, as Terry Erwin has shown, can be collected by the thousands, there are only a couple of hundred entomologists worldwide with the knowledge to differentiate them.

Once we find new species, of course we have to name them. As the numbers rise, that's getting to be a challenge all by itself.

Erwin, for example, has an estimated 2,000 species of ground beetles in one genus, Agra, waiting to be named. So far, he has named one *Agra vation* and one *Agra phobia*. A species with unusually large feet was named *Agra sasquatch*. A beetle in another genus was named *Pericompsus bilbo*, for J. R. R. Tolkien's beloved Hobbit character. Erwin explained the beetle was "short, fat, and had hairy feet." Yet another new beetle was christened *Agra katewinsletae*: I don't know what resemblance Erwin may have seen to the heroine of the film *Titanic*, and it may be best not to ask.

Erwin is not alone in the game of inventing unusual names for new species. In 1987, an entomologist who fancied the rock group The Grateful Dead named a fly *Dicrotendipes thanato-gratus*. The specific name combines the Greek word for "dead" and the Latin word for "grateful." A spider which sucked out its victims' bodily fluids was described in 1995 as *Draculoides bramstokeri*. Entomologist Arnold Menke named a new wasp genus after the thrill of discovery, calling it Aha. He then named a new species *Aha ha*. There is another wasp named *Heerz tooya*, while *Ba humbugi* is a snail.

A new woodland beetle from the northeastern United States received an intriguing scientific name in 1987. Collected eighty-five years before but misidentified as the known species *Platynus decentis*, the new bug was christened *Platynus indecentis*. Entomologist Kip Will explained that the animal's ability to exist for so long without being properly classified "seemed positively indecent."

Naming a species for a discoverer is common, but sometimes that process also takes on a humorous flair. Marine biologist John McCosker wrote, "I once named a gnarly, toothy eel for an occasionally irascible scientist; the favor was returned by colleagues who befitted morays and slime eels with *mccoskeri*."

This name game may reflect the perception that taxonomy is a field in need of some livening up. Or it may just be that, faced with millions of new species and a limited number of Latin and Greek words, science had to get creative sooner or later.

SOMETHING NEW OUT OF AFRICA

South America may lead the world's continents in bird species, but Africa has plenty to boast about concerning avian diversity. Well over 2,000 species inhabit the deserts, mountains, savannahs, forests, and valleys of the continent, and new birds are added to this list all the time.

One of the new birds described in 1995 was a species of night-jar from Ethiopia. The story of its discovery is a unique tale. The species was identified based on a single specimen, or rather on what could be salvaged after said specimen had been run over by a truck. A park warden and four visiting British scientists dug out the remains of the bird. Feathers on the only surviving part, a wing, showed markings unlike those of known species. The discovery was named *Caprimulgus solala*, or the Nechisar nightjar. (The Nechisar National Park was the site of the discovery, and the species name, solala, means "only a wing.")

In other avian news from this region, a new songbird was described from the African island of Madagascar in 1996. It received the interesting name of "cryptic warbler" (*Cryptosylvicola randrianasoloi*).

Another new African bird, a robin, was described from the Central African Republic in 1998. Pamela Beresford, a graduate student working with the American Museum of Natural History, collected the bird in late 1996. It took two years to complete the process of comparing it with other specimens and establishing its uniqueness. The new species was described as olive-brown with a yellow belly and yellowish-red plumage on its throat and upper breast.

According to Dr. Phil Hockey of the University of Cape Town in South Africa, "Ten years or so ago, ornithologists were saying that by now all bird species would be known. But today new species are popping up all over the place." Hockey stated that he knew of two more species awaiting formal descriptions, plus many other potential new species which had been seen but not yet caught or classified.

In the last half-century, Hockey said, forty-seven new species of birds have been snared in the wild. That number, impressive as it is, doesn't even include the larger number of new species described from restudy and reclassification of existing specimens. Another ornithologist, Rolf A. de By, counted fifty-nine species described from 1993 through 1999. Another twelve had been reported but either remain unconfirmed or have been challenged as inaccurate.

As the reader will notice from this example and the previous essays on birds, there is considerable variation in the counting of new species. Factors involved include what publications are deemed acceptable by whichever authority is doing the tallying, what descriptions are considered adequately documented, and whether reclassifications of existing specimens are included.

Whatever numbers are used, it's obvious there are many interesting new species for ornithologists to study. There are probably just as many more species we have yet to find at all.

THE WEIRDEST WORMS

In 1997, one of the weirdest animals on Earth—a worm living in masses of frozen methane—was discovered in the Gulf of Mexico. Biologist Charles "Chuck" Fisher of Pennsylvania State University and submersible pilot Phil Santos spotted the "wall of worms" from a submersible over 2,000 feet down.

The worms, christened *Hesiocaeca methanicola*, are up to two inches long. They have stubby pink bodies, pink feet shaped like paddles, and tufts of white bristles all over. They live in large numbers on brownish-yellow chunks of methane hydrate, some six feet across, which protrude into the water from huge masses of the methane-and-water composition frozen under the sea bed.

It's not yet clear how the worms obtain nourishment. They don't appear to live in symbiosis with any sort of microbes which could consume the methane hydrate. One theory is they eat bacteria which live inside the holes the worms hollow out in the ice.

The same research team had already made a major discovery in the same region in the 1980s, when new species were found clustered around brine pools on the ocean floor. These pools, formed by minerals seeping up through the bottom, may be thought of as large blobs of saltier, denser water clinging to the seafloor. They are surrounded by colonies of mussels and filled with water so dense a submersible cannot dive into it. Near these pools, in water 1,700 feet deep, are tubeworms of the genus Lamellibrachia. The worms are ten feet long and only half an inch thick. Like the larger tubeworms of the phylum Vestimentifera, these harbor chemoautotrophic sulfide-oxidizing bacteria, although the worms' exact mechanism for obtaining nourishment is still being investigated. The worms cluster in millions, looking like an undersea forest of strange plants. It is worth noting the fact that an animal whose colonies cover acres of the seafloor was not found until so recently.

In February 2000, it was announced that four years of measuring the growth of these tube worms had produced startling results. Judging by their size and growth rate, measured by marking individual tubes and returning year after year for measurements, the worms could be 250 years old. The Penn State scientists reported this age as a record for the animal kingdom, although some specialists believe giant sponges live longer.

We are still learning how strange life can be.

THE NEWEST WHALES

If any order of sea creatures should have been completely catalogued by now, it is the whales. After all, these mammals are generally large and must come to the surface periodically, announcing their presence with a conspicuous spout of exhaled air. Despite this, some members of the order Cetacea have avoided science until recently. Most of these belong to the rarely-seen family called the beaked whales.

Until recently, the Indopacific beaked whale (*Mesoplodon pacificus* or *Indopacetus pacificus*) was known only from two skulls washed ashore thousands of miles and seventy-three years apart. Dr. Lyall Watson, in his 1981 book, *Sea Guide to Whales of the World*, suggested a large pod of beaked whales photographed near Christmas Island might belong to this species, which is also called Longman's beaked whale. He made the same suggestion concerning brown whales reported in the Gulf of Aden by Captain Willem F. J. Morzer-Bruyuns, although that witness was certain he was seeing an unknown type of killer whale. Numerous other possible sightings of Longman's beaked whale, such as a report of two unidentified grayish whales seen near the Seychelles in 1980, were recorded, but no one was certain which ones—if any—referred to this enigma of the seas.

All that changed in 2002. An odd beaked whale beached on July 26 in Japan, but no one thought much of it initially. The carcass was photographed, then buried. When a cetologist saw the pictures, he scrambled to get the thing disinterred as quickly as possible. It was the first example of Longman's beaked whale ever recovered intact. In an odd coincidence, a second specimen identified as *I. pacificus* drifted ashore in South Africa the following month. (Two old South African specimens, which had been identified as other species, were then re-examined and were reported to be Longman's whale as well.) Until this point, cetologists knew nothing of the animal's appearance (it's predominantly grayish brown, with the head often appearing darker and sporting some small white side markings) and were unsure of its size (about twenty feet in length).

The smallest beaked whale is the Peruvian, or Lesser, beaked whale. Scientists had no inkling of its existence until 1976, when Dr. James Mead found its decaying skull on a beach in Peru.

By the time Mead formally published his description of *Mesoplodon peruvianus* in 1991, Peruvian scientists and fishermen had helped him assemble a total of eleven specimens. All were found either washed up on shore or trapped in fishing nets.

The adult Peruvian beaked whale is normally about eleven feet long. It is mainly dark gray, with a paler gray underside. It has a small dorsal fin set well back on the body. While all known mesoplodonts have such dorsal fins, there are differences in shape which help distinguish the different species. In some beaked whales, like the Peruvian, the fin is a near-perfect equilateral triangle with a straight trailing edge. In others, such as True's beaked whale (*M. mirus*), the trailing edge is concave, so the fin is more falcate or sickle-shaped.

It turned out this whale has a wider distribution than originally thought. Other specimens have since been found stranded in Mexico near Baja California and on the island of Espiritu Santo in the eastern tropical Pacific. There are still few recorded observations of the living animal, although pods of two or three have been seen swimming together.

In 1995, four cetologists published the results of their study of a single calvarium (upper skull) found on the beach of Robinson Crusoe Island off Chile in 1986. Julio Reyes and his colleagues proclaimed they had identified another new species of beaked whale. Bahamonde's beaked whale (*Mesoplodon bahamondi*) was distinguished principally by an unusually short and broad rostrum (snout).

The discoverers suggested Bahamonde's whale could represent the mysterious Mesoplodon "Species A," an unidentified beaked whale reported and photographed in the Eastern Tropical Pacific region. The overall length of *M. bahamondi* is estimated at sixteen to eighteen feet, which is an approximate match to these sightings. British paleobiologist Darren Naish, who makes a specialty of studying unusual cetaceans, cautioned that, "Glimpses of the head of Species A do not reveal the very abrupt rostrum that seems to be diagnostic for *M. bahamondi*, so they are probably not the same."

As things turned out, Naish was right. Bahamonde's beaked whale was not Species A—but it was identical to another mystery species. In a paper published in 2002, a group of cetologists demonstrated that *M. bahamondi*, while a valid species, was a resurrection of a species described in 1874 but generally forgotten. *Mesoplodon traversii* was restored to its rightful place in the genus after 128 years, while *M. bahamondi* was reduced, in taxonomic parlance, to the status of a junior synonym. This does not diminish the importance of the work by Reyes and company. It's significant any time a genuine new whale goes into the books—whether it's brand new or just a case of science saying hello to a long-forgotten discovery.

The beaked whales still had some surprises in store for science. One of the peculiarities of this group of cetaceans is that, while experts like Dr. Merel Dalebout of Dalhousie University in Halifax, Nova Scotia estimate they have been reproductively separated for perhaps three million years, their morphology hasn't changed nearly as much as their genetics. It's common for genetic change to lead morphological change, resulting in species that are distinct but still look similar, but the beaked whales have taken this principle to an extreme. Not only do many of the twenty-one known species look similar in life, requiring an expert to distinguish them, but even when an animal is beached it can be mistaken for another species.

That was the case with Perrin's beaked whale, *Mesoplodon perrini*. There have been many sightings of beaked whales which puzzled observers. For example, Dr. Karin Forney of the Southwest Fisheries Science Center spotted one off the coast of Oregon in 1996. The animal was brownish-gray and bore some resemblance to Hector's beaked whale (*M. hectori.*) However, some important details failed to match up. Most adult male beaked whales have two or more teeth in the lower jaw, which in many species are erupted (that is, are visible even when the jaw is closed.) The shape and placement of these teeth is a major criterion for classifying these enigmatic cetaceans. In Forney's whale, the visible teeth were not close to the beak tip as is normal in *M. hectori*.

As it turned out, the similarity to *M. hectori* was significant. Between 1975 and 1997, four beaked whales stranded on the coast of California were initially identified as Hector's beaked whales. Dr. Dalebout and her associates, in surveying DNA samples from numerous mesoplodonts, found these four didn't fit well with *M. hectori*. Neither did a fifth California specimen, which had been identified as Cuvier's beaked whale. In 2002, Dalebout, along with four of her colleagues, published the discovery of *Mesoplodon perrini*. When Karin Forney saw the description, she knew what she had observed in 1996. It was indeed, at the time she'd seen it, an undescribed whale.

In November 2003, a team including biologists Tadasu Yamada of Tokyo's National Science Museum and Shiro Wada of the National Research Institute of Fisheries Science made one of the most startling claims to hit cetology in a long time. Cetologists had long been divided over whether one of the smaller baleen whales, Bryde's whale, was a single species or might be two. Wada's team reinforced some earlier studies by announcing that, based on DNA evidence, Bryde's whale was indeed two species, *Balaenoptera brydei* (described 1913) and *B. edeni* (described 1878). (In an oddity arising from the rules of taxonomic priorities, the older scientific name, *B. edeni*, was the accepted as the proper term, even though Bryde's remained the common name.)

More importantly, according to studies of specimens taken by Japanese whalers in the 1970s, a third species had been missed entirely. Whales from the Solomon Sea and the eastern Indian Ocean which had been identified as unusually small fin whales were in fact an unsuspected species. Another such whale had stranded on a Japanese island in 1998, providing a complete carcass for study, and this happenstance allowed the scientists to confirm their work with the older remains.

The new species, *Balaenoptera omurai*, shows differences in appearance, skeletal structure, and DNA from fin whales and all others. *B. omurai* (named for a late Japanese cetologist) has an adult body length under forty feet. Compared to related whales, it has fewer baleen plates in its mouth. The skull is also different, relatively flat and broad. Wada made a point of giving credit to Japan's controversial program of "research whaling," which uses a loophole in International Whaling Commission (IWC) regulations to take several hundred of the smaller baleen whales every year. "Without that program, we would not have made this discovery," he said.

James Mead and others cautioned that, due to the variability within species of whales, it wasn't a given that Wada and Yamada were right. More thorough comparison with many other fin whale specimens will be needed before cetologists are certain. Yamada, though, has no doubts. ''Can you imagine?" he asked. "An animal of more than ten meters was unknown to us even in the 21st century." So far, the Japanese team has found ten examples of what it believes to be *B. omurai* in the world's museums. The Japanese announcement will no doubt set off a scramble to examine whale specimens all over the world—and to take a closer look at those still cruising the planet's oceans.

THE ANIMALS FROM HELL

Life on Earth has seeped into every conceivable niche. Animals have conquered the skies, the seas, and almost every square yard of land. What we have learned in the last two decades, though, is that life has moved into environments more difficult and inhospitable than scientists once thought possible.

The discovery of new worlds of life began in 1977, when Dr. Robert Ballard found the first seafloor geothermal vents near the Galapagos Islands. Nine thousand feet beneath the surface, stunned explorers on board the submersible *Alvin* found themselves looking at a scene from some alien planet. In the hot, mineral-laden water gushing from the seafloor was an entire ecosystem no one had dreamed existed. There were tubeworms up to eight feet long, so strange that describing them required creation of a new phylum. White clams, yellow mussels, and ghost-white crabs clustered around the vent kingdom, and weird-looking pink fish patrolled the edges of this island of life.

Biologists had a hard time accepting this at all until photographic proof was obtained. Holger Jannasch of the Woods Hole institute recalled receiving a call via a radio link from the Galapagos expedition. "The chief scientist…said he had discovered big clams and tube worms, and I simply didn't believe it. He was a geologist, after all."

How could life exist in this situation? Photosynthesis, the chemical engine driving all previously known life, is dependent on sunlight, and the last vestige of light vanishes about 800 feet below the ocean surface. It turned out the thick mats of bacteria growing around the vents were the foundation of the colony. Most animal species of the vent colonies—tubeworms, clams, and mussels—lived in symbiosis with the bacteria, and other species lived off these animals. The tubeworms of the phylum Vestimentifera, for example, are hosts to sulfur-oxidizing bacteria. These provide carbon for energy to the tubeworm. In exchange, the worm supplies hydrogen sulfide (toxic to almost every other form of animal life), oxygen, and carbon dioxide absorbed from the water by the plume-like red gills which protrude from the end of the tube.

By 1994, geothermal vent colonies had produced 300 new species of animals. In addition to the new phylum, ninety new genera and twenty new families have been created to house all these discoveries. In the mid-Atlantic, scientists found vents they called "black smokers." Minerals gushing from the vents built up "chimneys" rising from the ocean floor, clustered with transparent tubeworms, eyeless shrimp with light-sensing organs on their backs, and other bizarre animals. The water at one chimney was measured at 572 degrees Fahrenheit.

One vent colony inhabitant has been named "the Pompeii worm." In the outflow of vents on the East Pacific Rise, off Costa Rica, *Alvinella pompejana* tolerates both higher temperatures and greater temperature variations than any other multicellular creature on Earth. While the worm's gills stick out of one end of its tube home into relatively cool vent water at 72 degrees Fahrenheit, the creature's body is permanently enveloped in water measured at 176 degrees.

Researchers from the University of Delaware have studied the worm in its environment using the ubiquitous *Alvin* (hence the name of the creature's genus.) The study team suggested certain bacteria living on the worm, which envelop it in a fuzzy-looking coating, may possess enzymes which help the host animal cope with extreme temperatures. "It may be that the bacteria are insulating the worms in some way," speculated molecular biologist Craig Cary, the study director. Other animals (including humans) can survive such high temperatures for short periods, but the limit for prolonged exposure was previously thought to belong to an ant of the Sahara Desert which forages at a brain-frying 131 degrees.

Some researchers have questioned the accuracy of Cary's temperature measurements, noting that worms do not survive in the laboratory when subjected to such extremes. Even if the temperatures are not precise, though, the worms' adaptability remains impressive.

Discoveries from the vent colonies just keep coming. Timothy Shank, a marine biologist at Woods Hole Oceanographic Institution, estimates the vents have produced, on average, one new species every week and a half since 1979, and "we're still on the tip of the iceberg."

Then there are the microbes of the "deep biosphere." While most of these are bacteria or the more primitive archaeans, not animals, it is thought-provoking to realize we have only recently discovered an environment which might hold a greater biomass than the entire surface world. In the 1970s, the first enterprising biologists began culturing microbes in samples taken from hot springs, oil wells, and similar passages. Far below the presumed limit for living organisms —two miles down, in some cases—microbes living on minerals and respiring without oxygen have spread over huge areas. Some live in rock that is radioactive: others bask in temperatures above the boiling point of water. One researcher figured out that, if the pores and cracks in rock occupy three percent of the upper three miles of the Earth's crust, and microbes exist in one percent of that space, there is enough biomass to cover the planet's entire surface five feet deep.

In 1998, scientists were able to observe the way these two worlds —vent colonies and subterranean bacteria—came together to create a new outpost of life. Off the Oregon coast, a volcanic eruption ripped up the terrain, spread lava over the existing seafloor, and opened vents which poured out superheated water inhabited by high-temperature bacteria. The bacteria soon settled onto the floor and formed the living mats which are the basis of vent colony life. Tube worms and other inhabitants soon colonized the vent, and a new ecosystem was born.

Creatures living under the harshest conditions are often called "extremeophiles." What might be called an extreme extremeophile turned up in samples from a Pacific seafloor magma vent in 2003. "Strain 121" is one of the archaea, a domain of microbes lacking nuclei and believed to be the oldest type of organism on Earth. Strain 121 uses iron oxide from the vent to metabolize organic molecules in a way similar to how most organisms use oxygen. Strain 121 is an incredibly hardy microbe which does fine when put in a pressurized autoclave, used to sterilize medical instruments, and baked at 250 degrees F. Microbiologist Derek R. Lovley, who described the species, commented, "It has been the dogma in microbiology for 120 years that that temperature would kill any living organism." Strain 121 not only lives at that temperature, it multiplies. Lovley had to raise the temperature to 266 degrees to get the microbe to stop breeding, and even then it didn't die.

As a final note on this subject, a team from the University of Innsbruck reported in August 2000 that bacteria belonging to several, yet-unidentified species were living and reproducing in clouds. Water droplets captured in clouds that passed over Mount Sonnblick, near Salzburg, were analyzed and revealed to be active microbial habitats despite the conditions they offered—high levels of ultraviolet radiation, below-freezing temperatures, and minimal nutrients.

Biologists can only shake their heads in wonder at the adaptability of the still-mysterious stuff we call life. To modify a quote from Shakespeare, "There are more life forms on heaven and earth than are dreamt of in your philosophy.

SECTION II: IN THE SHADOWS OF EXTINCTION

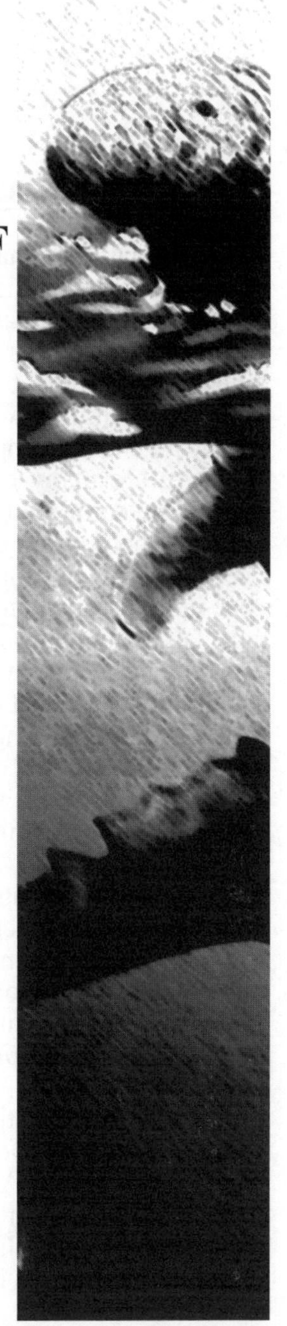

INTRODUCTION: SECTION II

Recent discoveries have reminded us that a species can go unobserved for years (sometimes decades) and still not be extinct. One of the most exciting areas of zoology is the quest to rediscover presumably lost animals. For some, that quest becomes a near obsession, as when conservationist Peter Zahler doggedly searched the mountains of Pakistan until the world's largest squirrel—long presumed extinct—was practically dropped in his lap.

There are many other recent examples. Take Lowe's servaline genet, a predator related to the mongooses. *Genetta servalina loweia* is a sleek-bodied, spotted mammal about three feet long. This animal was collected once, in 1932, and then vanished from the view of science until it was photographed in Tanzania seventy years later. Even very large animals can pull disappearing acts under the right conditions. Recent rediscoveries include the Vietnamese population of the Javan rhinoceros, not confirmed since the 1960s, and the giant black sable antelope of Angola, a subspecies written off for forty years before it was located once again in 2002.

Leaving aside the mammals, there are examples from every corner of the animal kingdom. Take owls. The Indian forest owlet, the Madagascar red owl, and the Congo bay owl were found in the last few years after going missing for a long time—113 years, in the owlet's case. The nineteen-inch lizard *Gallotia gomera* of the Canary islands was known only from remains estimated to be 500 years old until the discovery of living specimens was announced in March 2000. One of this animal's closest relatives was the Hierro giant lizard (*Gallotia simoni*), which was also thought extinct until its rediscovery in 1975.

Those survivors still awaiting discovery may include some spectacular creatures. Does the thylacine still haunt the forests of Tasmania—or even Australia? Has the Eastern cougar slipped back into its old haunts in eastern North America? Could the Caribbean monk seal still be frolicking in the warm waters of its habitat? Not all of these creatures will be rediscovered, but some likely will be.

Steller's sea cow, believed hunted to extinction in the 18th century.

*That's enough of a reason to keep up the scientific detective vork in the hopes of restoring a precious piece of the natural world.

The true bombshell, though, came on April 28, 2005, with the nnouncement that the ivory-billed woodpecker (*Campephilus prinipalis*) had been rediscovered in the United States after a long and eemingly permanent absence.

The world's largest woodpecker, the magnificent black, white, nd red ivory-bill was nicknamed "the Lord God bird" after the xclamation many people uttered when seeing one. This species' xistence had not been confirmed in the United States since the late 940s, although scattered reports, like a sighting claimed by ornithologist John V. Dennis in eastern Texas in 1966, kept interest alive.

* PUBLISHER'S ERRATA

The dodo bird, symbol of extinction.

When, around 1991, sightings also ceased in the bird's last known refuge in Cuba, most ornithologists sadly wrote it off for good.

An impressive sighting by graduate student David Kulivan in Louisiana in 1999 sparked a revival of interest, setting off a major search which unfortunately failed. Five years later, though, in the Big Woods of eastern Arkansas, kayaker Gene Sparling watched a strikingly large woodpecker whose wings bore the distinctive white trailing edges of the ivory bill land only twenty yards away. Sparling's account was convincing enough to bring Cornell University's Tim Gallagher to the area, along with wildlife photographer Bobby Harrison. On February 27, 2004, an ivory-bill appeared in flight just ahead of the men's canoe, then veered into the trees. A massive, expensive, and highly clandestine search resulted in five more sightings and four crucial seconds of videotape. Veteran ornithologists, some of whom had searched for decades, wept as they watched the tape. Efforts are underway to expand the protected areas in the Big Woods and otherwise give the ivory-bill the best possible chance to make its amazing comeback permanent.

To keep up the search for "lost" species is not just a scientific desire, but a moral imperative. Estimates of the number of species becoming extinct each year vary from one (the average rate for documented extinctions of known species) to an extreme of 40,000. Even given that the latter figure is an unverified worst-case estimate, the situation is serious at best.

Even among our closest relatives, the primates, the World Conservation Union estimates 130 of the 600-odd known species are endangered. Several species and subspecies are in a frightening twilight stage, where we simply don't know whether they still exist at all. In September 2002, a subspecies of large African monkey, Miss Waldron's Red Colobus, was declared extinct. This marked the first known primate extinction in modern times. (The pronouncement of the species' death might have been premature—John Oates of New York University, coauthor of the paper declaring the animal extinct, now thinks there's a tantalizing chance he may have been wrong. The bad news is the new evidence came in the form of three kills by African hunters, and any lingering population is a hair's-breadth from vanishing for good.)

The International Union for the Conservation of Nature (IUCN) regularly puts out a Red List of threatened species worldwide. The 2003 edition is not a heartening document. The number of mammal species labeled "Critically Endangered" rose from 169 on the 1996 list to 184, as birds climbed from 168 to 182 and reptiles from 41 to 57. The number of mammals in one of the Red List categories of concern (Critically Endangered, Endangered, or Vulnerable), was 1130 in 2003, almost a quarter of the total mammal species examined.

Examples like this are fueling a growing consensus that to drive another species unnecessarily into extinction is not acceptable if we are to call ourselves "human." One need not agree with the radical "terrists" who claim humans are "a cancer on the planet" to recognize that the loss of species is a tragedy which must be prevented whenever possible.

In 1906, William Beebe expressed the thinking behind this concern in these words:

The beauty and genius of a work of art may be reconceived, though its first material expression be destroyed; a vanished harmony may yet again inspire the composer; but when the last individual of a race of living things breathes no more, another heaven and another earth must pass before such a one can be again.

THE SEARCH FOR THE WOOLLY SQUIRREL

The giant woolly flying squirrel (*Eupetaurus cinereus*) is one of the most amazing mammals most people have never heard of. A resident of the Himalayan region, the woolly may be four feet long, almost half of the length being tail. It is not just the largest flying squirrel but the largest squirrel of any kind in the world.

The woolly squirrel was discovered in 1888, but almost completely vanished after that. Apparently only one was ever kept in captivity, and few specimens reached museums. By the 1990s, the animal hadn't been seen in decades. It was widely considered extinct.

A brief item on this "missing" animal in the authoritative guide *Walker's Mammals of the World* fired the imagination of Peter Zahler, a New York writer and conservationist. At considerable personal expense, Zahler and his friend Chantal Dietemann traveled to Pakistan in 1992 and spent two fruitless months searching for the animal. The incredibly determined Zahler obtained a WWF grant enabling the twosome to return in 1994. They were nearing the end of this second expedition when Dietemann found the first sign of hope—a squirrel's front leg, apparently dropped by a predator. Even this clue didn't lead anywhere until two local men approached them, saying they knew where to catch a woolly squirrel. Zahler was skeptical, but, a mere six hours later, the Pakistanis dropped a bag containing a live specimen at his feet. Zahler was so stunned he didn't know how to react. Finally he managed to say the obvious: "It's a woolly flying squirrel."

The capture site was a cave in the Sai Valley, at an altitude of almost 10,000 feet. The Americans examined and photographed the five-pound creature, then released it and returned to the U.S. with their news. Mammologist Charles Woods, an expert on Pakistani wildlife, was "flabbergasted," noting, "I've worked all through there ...We've really scoured the area and never seen it. This is simply marvelous."

Part of the squirrel's ability to disappear is due to its choice of habitat. Given the universal predilection of other flying squirrels for homes in trees, one would hardly think to look for one living among caves and boulders. That, however, is where the woolly resides, using its "wing" membranes to jump from one rock to another. Another point in its favor is the remoteness and ruggedness of the terrain it prefers. "There could be a thousand of them twenty feet over my head, and I'd never know it," Zahler said.

The woolly is a bit strange in appearance. Its body size and crouching posture tend to make it look more like a gray raccoon than a squirrel. The long tail is luxuriant and foxlike.

Zahler returned to Pakistan in 1995. Again he caught one squirrel, which he held overnight and then released. He planned future efforts to radio-collar some specimens and learn more about their habits. So little is known today about the species that even its diet is uncertain. The 1987 *Collins Guide* to rare mammals said the species was "believed to feed on lichens and moss." Zahler reported it apparently eats spruce buds. Either way, this explains the failure of his initial efforts to lure it into traps using honey, grain, and nuts as bait.

Zahler's goal now is to secure official protection for the animal. While there is no way to know the squirrel's population, it certainly isn't common. But at least, thanks to Zahler, we know it's still there.

WHO'S THAT OWL?

The history of zoology is replete with tales of animals which have been collected one time and never seen again. These type specimens often come to rest in obscure corners of museums, sometimes to be forgotten for decades or even centuries.

Occasionally, though, an animal is rescued from this scientific purgatory. The Congo bay owl (*Phodilus prioginei*), also known as the Itombwe owl, has been known since 1951 from a single specimen collected in the Itombwe Massif of what is now the Democratic Republic of the Congo (known until 1997 as Zaire). It had never been reported before this, and there were no confirmed sightings for forty-five years after the original incident. In fact, there was only one unconfirmed sighting, which came from the neighboring nation of Burundi in the 1970s. Due to the total lack of information, the International Council for Bird Preservation (ICBP) listed the species' status only as "indeterminate."

It took until 1996, but the little African owl was found again. In that year, a team from the Wildlife Conservation Society and the Zaire Institute for Nature Conservation penetrated the Massif, a wild region in the eastern part of the nation, to survey its animal life. Here, in the largest undisturbed montane environment left on the African continent, they captured a female Congo bay owl.

This isn't the only recent rediscovery of a member of the owl family. Another success from 1996 was the confirmation of the continued existence of the Madagascar red owl (*Tyto soumagnei*). A press release from Conservation International stated this species had not been seen since the 1930s, although the ICBP recorded a lone sighting report in 1973.

In 1997, an equally rare and mysterious bird was rediscovered. The Indian forest owlet (*Athene blewitti*) was photographed after a 113-year absence by American ornithologists Ben King, Pamela Rasmussen, and David Abbott northeast of Bombay.

The owlet, a brownish, eight-inch-tall bird with disproportionately large feet and beak, was first recorded in the 1870s and had a very brief acquaintance with science. The last verified encounter with this species took place when an individual was caught in 1884. When modern ornithologists set out to look for it, the best illustration they had to go on was a painting made in 1891—which turned out to be inaccurate anyway.

As just mentioned, no specimens of the Indian forest owlet had been collected since 1884. However, a bizarre episode concerning this species took place in 1914. A British birder, Richard Meinertzhagen, produced a "new" stuffed specimen in that year. A modern re-examination of this case, though, performed by Rasmussen and fellow ornithologist Nigel Collar, established that Meinertzhagen's bird was an individual belonging to the British Museum. The specimen had been stolen and restuffed.

After the Meinertzhagen affair, nothing was heard of the species except a few doubtful reports and two photographs that turned out to show owls of other species. When the three Americans began their search, there were few clues to go on. The best they could do was check areas nearest to where long-ago captures had taken place. They were about two weeks into checking one forested area when Ben King pointed a finger and said quietly, "Look at that owlet."

Rasmussen stared at the bird, dropping her water bottle. She remembered, "My instant reaction was, 'this thing is going to fly, and I'm not going to be able to verify it.'" Fortunately, that was not what happened. The first *Athene blewitti* seen in over a century sat still for half an hour while the stunned ornithologists videotaped it. Now that it's truly been rediscovered, we can hope this enigmatic avian can be properly studied and conserved so it will not vanish again.

Owls are not the only birds to have turned up in this fashion. The list of long-lost birds which have been rediscovered is surprisingly extensive. It includes the Eskimo curlew, the Fiji petrel, the Bermuda petrel (presumed extinct for nearly 300 years!), and Fiji's long-legged warbler, the last found again in November 2003 after an absence of ninety-nine years. The indigo-winged parrot of Columbia was reported and classified in 1911. Then it was gone. This colorful green, blue, and red bird, *Hapalopsittica fuertesi*, was not spotted for the next nine decades. In 2002, ornithologist Jorge Velasquez searched the Andean jungles for the parrot. He found nothing in four months of searching. Then, out of the mists, like "a miracle from heaven," he recalled, a flock of fourteen birds appeared and alighted in the tree branches in sight of Velasquez and his companion, Alonso Quevedo. The indigo-winged parrot was back.

A large and spectacular example is the Madagascar serpent eagle (*Eutriorchis astur*). Presumed extinct since 1930, the eagle was reported sighted in 1988. Biologist Russell Thorstrom, of the conservation group The Peregrine Fund, saw one in 1993. He returned to the same area repeatedly, searching for something never before seen: a serpent eagle nest. He finally found one in November, 1997. He camped out for three months, obtaining the first-ever photographic record of this raptor raising its young. Biologists are searching for more nests and trying to develop a better understanding of how many eagles are left and where they live—vital steps toward saving the species.

MORE TREASURES FROM SOUTHEAST ASIA

In the last section, we saw how many new large mammals have turned up in the Vu Quang region. The same area has produced three major rediscoveries of presumably-extinct mammals as well.

The most recent and unexpected find was the rediscovery of Roosevelt's muntjac, a.k.a. Roosevelt's barking deer.

The story began in 1929, when a young male muntjac was collected in Laos. It was described as a new species, Muntiacus rooseveltorum. As decades passed without another specimen, the species was assumed to be extinct. It might have faded into history had not visiting scientists including zoologist George Schaller observed an unusual deer in January 1995 in a menagerie in the Laotian village of Lak Xao.

The Lak Xao specimen, with its dark coat, resembled the known "black muntjac" (*M. crinifrons*), but that species does not occur in Laos. Also, this animal was larger and had different cranial characteristics—similar, some of the visitors thought, to the almost-forgotten *M. rooseveltorum*.

Intrigued, Schaller and others obtained six full or partial skulls they were told belonged to the same type of deer. Comparing these to the type specimen, now in the Field Museum in Chicago, they pursued the matter to a final identification. The conclusion, published by George Amato, Schaller, and others, was that science had rediscovered Roosevelt's muntjac. Ironically, it turned out the animal seen on that January 1995 visit to the menagerie was apparently not Roosevelt's muntjac, but most likely a specimen of the newly discovered Truong Son muntjac (see Section I). That a false clue had led to an accurate conclusion is a reminder that science, like God, works in mysterious ways.

In their article on Roosevelt's muntjac in the *Journal of Mammology*, the authors note, "This study highlights the importance of continued field surveys in remote regions and the utility of diagnostic DNA characters in identifying species." Indeed, they were able to sort out a number of muntjac mysteries. Their DNA work confirmed MacKinnon's description of the Truong Son muntjac as a separate species, while suggesting one older description, *M. gongshanensis*, might not be valid.

The Vietnamese warty pig, *Sus bucculentis*, had been described in 1892 when Father Pierre-Marie Hende, a Jesuit missionary, literally drew the attention of science to the animal by publishing a sketch of its skull. (A surprising number of contributions to zoology have been made by missionaries, and even today some carry a

Bible in one hand and a field notebook in the other. The subject would make a fascinating book all by itself.)

Two skulls of the warty pig were eventually obtained from southern Vietnam. No live specimen was ever seen by a Western scientist. The species seemingly disappeared in the wild, and was long considered extinct or even invalid.

On the same WCS survey that produced the giant muntjac and Roosevelt's muntjac, George Schaller and Laotian scientist Khamkhoun Khounboline were told of a yellowish-furred pig with a long snout. After turning up a partial skull and a tissue sample in 1995 and locating one of the long-missing type skulls in Beijing in 1996, George Amato and Colin Groves matched the evidence to the 1892 description of *S. bucculentis*. The species may be extinct in Vietnam, since the rediscovery took place in Laos, farther north along the rugged Annamite mountain range.

Finally, we have the mainland population of the Javan rhinoceros. The species *Rhinoceros sondaicus* was presumed wiped out on the Asian mainland since the 1960s. It's almost absurd to think a 3000-pound rhinoceros went missing for thirty years, but it did. No one reported the species again until a 1998 sighting in Lam Dong province in Vietnam. In July 1999, the first photographic evidence, from Dong Nai province (some eighty miles north of Ho Chi Minh City) was obtained. Perhaps twenty of the animals—maybe fewer—live in Vietnam, adding to the precarious population of fifty to sixty in Indonesia. Steve Osofsky of the WWF calls the rhino "the most endangered mammal species in the world."

GILBERT'S POTOROO

Australia abounds with animals which have disappeared only to turn up again long after zoologists have written them off. Gilbert's potoroo (*Potorous gilberti*) is the latest example, and its recent rediscovery is one of the most surprising such cases on record.

Gilbert's potoroo is a miniature relative of the kangaroo family. (It is sometimes classified as *Potorous tridactylus gilberti*, a subspecies of the long-nosed potoroo.) The potoroos are saddled with one of those Dr. Seuss names created when an Aboriginal term is approximated by English speakers. Their other common name, rat-kangaroos, may not sound as silly, but isn't very flattering, either. Less than eighteen inches long, potoroos do somewhat resemble pudgy rats fitted with kangaroo-type hind legs.

This particular potoroo, with a distinctive black stripe on its face and a black tail, was known only from two specimens taken in the last century by John Gilbert in Western Australia. Zoologists assumed the creature was rare then and vanished for good shortly after. Most modern references flatly list Gilbert's potoroo as "extinct," although the 1991 edition of *Walker's Mammals of the World* noted unconfirmed sightings a decade earlier which might refer to this animal. Introduced foxes and habitat destruction were blamed for the elimination of the species, which lived mainly around streams and in swampy areas.

In December 1994, Environment Minister Kevin Minson announced the marsupial's reappearance. In a nature reserve some 250 miles south of Perth, five specimens of Gilbert's potoroo were live-trapped. The examples collected included two adult males, one juvenile male, and a female with a youngster in her pouch. The size and viability of the surviving population remains to be determined, but the species' continued existence gives us hope for other missing animals as well.

MONK SEAL SURVIVAL?

Is it possible the Caribbean monk seal (*Monachus tropicalis*), presumed extinct for decades, still survives?

In 1997, ninety-three Haitian and Jamaican fishermen were interviewed about local marine mammals. Twenty-one of these witnesses identified the monk seal as a living part of the native fauna. Of these men, sixteen said they'd seen at least one monk seal within the last two years.

That's not definitive, but it is encouraging, since this species has not been confirmed in a long time. The monk seal was the only pinniped endemic to the Caribbean, and the first New World mammal to rate a mention in the journals of Christopher Columbus. Columbus' men killed eight of the brown animals they called "sea wolves." At the time, these mammals were abundant. They gathered in herds of up to 100 and ranged as far east on the coast of South America as Guyana. They also grew to impressive sizes. Males could be up to seven feet long and weigh 400 pounds.

Exactly how monk seals came to inhabit the Caribbean is not clear. Their closest relations are the world's other two species of monk seals. Of these, one lives in the Mediterranean, the other in the eastern Pacific. This zoological puzzle was irrelevant to European settlers, who knew an easy source of food, oil, and hides when they saw one. By the end of the 19th century, the seals had almost been wiped out.

In 1911, the last large colony—about 200 seals living on small islands off Yucatan—was slaughtered. Since then, reliable records of the monk seal have been rare. A lone individual was killed near

Key West, Florida, in 1933. A small group of seals which lived near Jamaica was observed until the early 1950s, but vanished.

After that, the record becomes even murkier. A monk seal was spotted lying on the beach in Rockport, Texas, in 1957. Others were reported by fishermen off Belize in the 1960s and near Yucatan in the 1970s. Searches in 1973 and 1980 came up empty, although the Seal Conservation Society's Internet site mentions a sighting report from an unspecified location in 1984.

It's been suggested some reports of Caribbean monk seals could be caused by California sea lions (*Zalophus califonianus*), escaped or released from oceanic parks along Florida's Gulf Coast. (Interestingly, escaped California sea lions have also been suggested to account for reports of the Japanese sea lion, *Zalophus japonicus*. The last known population of this species vanished from the island of Takeshima sometime after 1951.)

California sea lions are normally darker than monk seals, but their size ranges overlap, and the two could certainly be confused at a distance. Other sightings may involve wayward members of other species, such as the hooded seal (*Cystophora cristata*) which recently stranded in the U.S. Virgin Islands, a thousand miles south of its usual range. Until scientists catch or photograph a monk seal in the Caribbean—or prove once and for all that none exist—the mystery and the hope will remain.

BIRDING IN INDONESIA

The year 1995 brought rejoicing among ornithologists when no fewer than three birds, all from the Indonesian region, were rediscovered in the same year. All had been feared extinct after years or decades without a sighting.

The first species involved was the intriguingly named invisible rail (*Habroptila wallacii*), an always-rare species known only from the island of Halmahera. This bird was not classed as definitely extinct, since there had been scattered sightings from the 1980s and early 1990s, but its status was precarious at best. No one was certain it existed until the species was tracked down and confirmed in 1995.

There was more concern about the fate of the Lompobattang flycatcher (*Ficedula bonthaina*), known only from a small area in the dense forests of the Lompobattang massif on the southwestern tip of Sulawesi. Up until about seventy years ago, it was apparently common in this locality, and ornithologists had no trouble finding specimens. After 1931, though, sightings of this bird ceased. The area was largely cleared for settlement, and the lands below the flycatcher's favored habitat (above 3,000 feet) are now heavily populated.

The third species, the cerulean paradise flycatcher, (*Eutrichomiyas rowleyi*) was only discovered in 1873. It lived in one place—the island of Sangihe, north of Sulawesi. The flycatcher, a spectacular bird with blue plumage and a long tail, was not seen again until one report in 1978, after which it disappeared once more. Since the species' discovery, most of the island's native vegetation was converted to coconut and nutmeg plantations. Expeditions after the 1978 sighting had failed to locate the bird, and the IUCN listed the species' status as Critical.

The rediscovery of the flycatcher was made in 1995, then confirmed in 1999 in a new report by the conservation group Action Sampiri. According to British ornithologist Jim Wardill, his scientific team had identified at least twenty-two individual birds on Sangihe and thought the population might be over a hundred. The survey was done in October 1998, but the results had been kept secret until official protection for the bird could be secured. Wardill cautioned the bird was still seriously threatened by habitat destruction. "It's a miracle it's still there," he said. These days, conservationists need all the miracles they can find.

A final story from this region concerns the Chinese crested tern, *Sterna bernsteini*. This species once ranged from Indonesia north along the Asian coast, but the last specimens collected were from Shandong, China, in 1937. Since then, there were occasional reports from locations as far apart as Indonesia, China, Malaysia, and the Philippines. The species was pressured by human development, habitat destruction, pollution, and egg poaching. The last definite sighting was of a small flock in Thailand in 1980. After that, ornithologists wondered if the bird was alive anywhere.

Fortunately, it was. In July 2000, photographer Liang Chieh-teh found at least four active nests on the tiny island of Matsu, off Taiwan. Taiwan's Council of Agriculture announced the location was being kept secret, but photographs were published showing the black-and-white birds, with their distinctive black-tipped beaks, guarding their nests. One ornithologist, Dr. Ian Nesbitt, declared, "This is a very exciting discovery. The Chinese crested tern is one of the least known and possibly the rarest seabird in the world."

LEMUR RESURRECTION

The pygmy mouse lemur, *Microcebus myoxinus*, was described in 1852. The smallest primate in the world, this reddish-brown animal is marked by a white stripe down its face, a creamy underside, and a dark stripe down its back. It has large round eyes and weighs just over an ounce. For a long time following its discovery, though, no one paid the species any attention. Indeed, its description was generally written off as an error based on a misidentification of an immature specimen or a variation of the gray mouse lemur, *M. murinus*.

A tiny lemur caught in the Kirindy forest of western Madagascar in 1992 began the work to set the story straight. Two researchers from the German Primate Centre, Jutta Schmid and Peter Kappeler, determined the specimen was an adult and was not a gray mouse lemur. Among other things, the pygmy mouse lemur has shorter ears, a longer tail, and a lighter build than its cousin. Finding and examining more animals of the same type, Schmid and Kappeler showed the 1852 description had been correct after all, and the pygmy was restored to its rightful place in the primate world.

IS THE DEER STILL THERE?

The members of the family Cervidae, the deer, have generally done very well in surviving the expansion of humanity. Some species, like Pere David's deer (*Elaphurus davidianus*, another species originally found by a missionary), had close calls, but almost all are still with us.

The case of Dawson's caribou is one of those tales of animals whose existence was almost ghostly, as if the species was never really there. This animal had the smallest range of any known deer, being reported only from the northwestern part of Graham Island, one of Canada's Queen Charlotte Islands. It's not known when caribou came to this island, presumably by swimming, but they had been there long enough for some significant adaptations. Dawson's caribou was smaller and paler than any other caribou, standing perhaps three feet high at the shoulder. It bore antlers that were small and undeveloped, and the hoofs were broad for support in boggy terrain. Its numbers could never have been large, and the population was likely below the minimum of 50 individuals often cited as necessary for survival. (This number comes from the "50/500 rule," a postulate accepted by some biologists that says a reproductively isolated population, to remain viable—that is, to avoid the dangers of inbreeding and genetic drift—requires at least 50 individuals in the short term and 500 in the long term. The minimum numbers are lower in captive populations, where breeding can be selectively managed.)

Described only in 1882, Dawson's caribou was only seen a few times, and five specimens were killed. That, no doubt, was a major dent in the population. Some authorities date the last sighting to be around 1908, while others think Dawson's caribou may have lingered sometime into the 1930s. Some think the animal was a species, *Rangifer dawsoni*, although recent analysis indicates that it was a subspecies, *Rangifer tarandus dawsoni*, instead.

Whatever it was, Dawson's caribou lingered only a moment at the edge of humanity's consciousness, and then it was gone. In a bizarre postscript, it was reported in 1995 that the type specimen, which still survived in mounted form at the Royal British Columbia Museum in Victoria, British Columbia, was harnessed to Santa Claus' sleigh in a Christmas display. (To be extinct is bad enough—to be humiliated afterwards is unconscionable.)

Only one full species of deer has been classified as extinct in modern history. This is Schomburgk's deer (*Cervus schomburgki*), last seen in its habitat in Thailand in 1932.

This animal was never well known. No Westerner ever saw one in the wild, although a live specimen reached a French menagerie in 1867. Schomburgk's deer has impressive antlers, with multiple tines making up complex branches. In China, these antlers were believed to have medicinal properties. It was the demand for antlers, combined with drainage of the deer's preferred swampy habitat for agriculture, which drove this always-rare species toward extinction. The deer vanished so quickly that today there is only one mounted specimen in the world.

After the last known example of Schomburgk's deer was killed near the town of Savoy, nothing was heard of the animal for nearly six decades. By 1991, it was universally classed as extinct.

In that year, however, a United Nations agronomist found a set of this deer's distinctive antlers for sale in a shop in Laos. According to the shop owner, the specimen came from a remote, forested region of that country. Occasional reports of what might be a remnant population trickled in after that discovery, although there has been no more hard evidence. The October 2004 version of the IUCN Red List stated the animal was considered extinct, but added, "The matter has been referred to the relevant Specialist Group for a decision."

So far, that is the end of the trail. There appears to be a slight chance, though, that the story of recent mammal discoveries and rediscoveries in Southeast Asia includes one chapter that is yet to be written.

BRINGING BACK THE DEAD

Much of the work done in conservation and in cryptozoology involves searching for surviving members of species presumed extinct. Inevitably, that raises the question, "What if a species really is extinct? Have we lost it for good?"

The answer to this fascinating riddle is: maybe yes, maybe no.

Take the quagga, for example. This close relative of the zebra was immediately distinguishable by its brownish body, set off by brown striping on a white head and neck (except in the case of one freak animal, which was, believe it or not, hairless and blue.) The quagga became extinct in 1883, when the last one died in the Amsterdam Zoo.

A hundred years later, efforts began to resurrect the quagga, or at least a reasonable facsimile. Through capture and selective breeding of zebras which show signs of quagga ancestry, a group of enthusiasts led by South African Museum taxidermist Reinhold Rau is trying to revive the animal.

The Quagga Breeding Project was founded in 1986, but its origins go back to 1983. When samples from a stuffed quagga Rau was remounting were sent to American geneticist Russell Higuchi, it was established that the quagga (originally classified as a species, *Equus quagga*) and Burchell's zebra (*E. burchelli*) were cospecific, with their distinct appearances representing only subspecific differences at most. This discovery gave Rau the inspiration to search for brownish or partially striped Burchell's zebras and collect them for crossbreeding designed to bring out the quagga characteristics.

If Rau does produce something resembling a quagga, will it be a quagga? The question is difficult to answer. The problem is that, based on the mitochondrial DNA Higuchi has been able to retrieve, there is no definite way to tell a quagga from a zebra. It's like deciding whether a Siamese cat with tabby ancestry is really a Siamese: they belong to the same species, so it's a matter of definitions.

Geneticist Oliver Ryder of the San Diego zoo, who received the first quagga tissue specimens from Rau and passed them on to Higuchi, doesn't think so. "Quaggas are extinct," he says flatly. "There's a difference between producing animals that have the appearance of the quagga and actually resurrecting quaggas." Rau disagrees: "A quagga is a quagga because of the way it looked, and if you produce animals that look that way then they are quaggas."

By 1998, eleven "reconstructed" quaggas had been released in South Africa's Karoo National Park. A "perfect" quagga—one replicating the most distinctive specimens known from museums—may take another thirty years to breed.

Similar back-breeding experiments have been undertaken in the past, most notably by the Heck brothers, two German zookeepers. Lutz and Heinz Heck spent decades re-creating the long-extinct wild cattle called aurochs (*Bos primigenius*) and the tarpan, the prehistoric wild horse of Europe. As with the quagga project, the premise was that the genes of these extinct animals still existed in dilute form as a result of interbreeding with related animals. If so, these genes should theoretically be recoverable.

Again, it is a matter of dispute whether the Hecks' "aurochs" and "tarpan" are "real" or just curiosities which only look like the originals. The animals eventually produced by the Hecks do breed true to type. Today's tarpan appears identical to its ancestor, including the bristly mane, unusually hard hooves, and gray coat.

This is actually only one of three efforts to breed back the tarpan. In Poland, Professor Tadeusz Vetulani began the first endeavor shortly after World War I. Using *konik polski* (Polish horses) known to have tarpan ancestry, he also had success—until his herd was nearly wiped out in World War II. A few scattered horses with tarpan characteristics were located after the war and taken to the city of Polpielno, where Dr. Magdalena Jaworowska began another program of back-breeding. The horses bred so far show many tarpan traits, although they have more refined, "modern" features as well.

Americans Gordon and Leanette Strobel have recreated a tarpanlike horse from an unlikely source. Using American mustangs, they have, in effect, reached back into this animal's past to recover tarpan genes carried by the original horses brought by Spanish conquistadors in the sixteenth century. Today, anyone who wants a unique horse can buy a tarpan lookalike from the Stroebels for $3,500.

Another back-breeding program was sparked in 1996, when an unusual lion was rescued after being abandoned by a bankrupt circus in Maputo, Mozambique. Giepie, as he was called, appeared to be a Barbary lion. The Barbary lion was a North African subspecies. It had once been wide-ranging and numerous (the Romans captured a large number to devour captives in the Coliseum) but had vanished, at least in pure form, about the 1920s. The Barbary subspecies was marked by large size and a luxuriant dark mane that continued down most of the animal's underside. Unfortunately, these features made the lion a favorite of trophy hunters.

Giepie was sent to a wildlife research center in Hoedspruit, South Africa. Publicity over his capture caused zoos and private owners to examine their own collections for lions with Barbary characteristics. In 1998, a Barbary-type lioness which had been discovered in captivity in Italy was sent to Hoedspruit with hopes of starting a breeding program. While there is some uncertainty over whether the Barbary lion had genetic differences from other lions—"A Barbary is based on its look," explained the Hoedspruit center's veterinarian—the prospect of seeing this majestic animal restored has attracted widespread interest.

So has the apparent rediscovery of another lion with a similar appearance. In the fall of 2000, two lions in at the zoo at Novosibirsk in Siberia were tentatively identified as being descendents of Cape lions. The Cape lion, with a thick black mane and black tips on its ears, disappeared from its South African range in the mid-1800s. The lions at Novosibirsk are reportedly descended from a single animal left decades ago by a circus.

A Japanese-led effort to do something far more spectacular has so far come to naught. Geneticist Kazufumi Goto and his colleagues recently completed an enormously difficult and ultimately fruitless expedition to Siberia in search of frozen mammoth sperm. If the complete sperm of a mammoth can be recovered, their thinking goes, it can be used in a surrogate elephant mother to produce a baby which is half percent mammoth and half percent elephant. Further breeding could eventually produce a nearly pure mammoth.

An even more speculative venture of Goto's is using mammoth DNA to skip the hybrid stage and create an actual clone. If a mammoth cell with a complete nucleus can be found, it could replace the nucleus in an elephant egg and develop into a clone. There is no guarantee—and, some scientists believe, almost no chance—of success, but Goto is taking the quest one step at a time. He's planning more trips to Siberia to search for new mammoth remains. "It may sound unbelievable, but there's no science to deny our idea," he says optimistically.

In 1999, the possibility of cloning the supposedly extinct thylacine, or Tasmanian tiger (*Thylacinus cynocephalus*), was raised by Michael Archer of the Australian Museum. The Museum has a baby thylacine preserved in alcohol (other extant specimens were kept in formalin, which destroys DNA). The surrogate mother issue was a major problem, since the much smaller Tasmanian devil (*Sarcophilus harrisi*) would somehow have to serve. Professor Ian Wilmut, who created the world-famous cloned sheep, Dolly, dismissed the project's chances of success as "extraordinarily unlikely." In 2005, the museum gave up after the DNA was found to be too degraded for use.

Another recent cloning prospect is a subspecies of Pyrenees mountain goat, *Capra pyrenaica pyrenaica*. The last living example of this subspecies, known locally as the bucardo, was crushed by a falling tree in Spain's Ordesa National Park in January 2000. A tissue sample was immediately frozen with cloning in mind. The cloning possibilities are being investigated by a company called Advanced Cell Technology, which has already gotten a cow pregnant with an egg from a rare type of wild cattle, the gaur (*Bos gaurus*).

In New Zealand, an officially sanctioned effort is already underway to clone the huia (*Heteralocha acuturostris*), a native bird last seen alive in 1907. This feat would be of considerable scientific interest because the huia was one of the most unusual birds known. It is the only species in which the male and female had very different types of beaks and reportedly fed cooperatively. The male used his short, stout bill to open up bark and wood to find insects, while the female used her longer, curved bill to probe crevices the male couldn't reach into.

In 1998, New Zealand scientists reported the third example of successful cloning of an adult mammal. This had some implications for cryptozoology because the mammal cloned was the last living specimen of a breed of cow. "Lady" was the last survivor of a herd living on sub-Antarctic Enderby Island. A unique breed of cattle, with short legs and the ability to live mainly on seaweed, developed in the island's harsh conditions. David Wells, leader of the science team, described this case as "proof of principle" that cloning could be used to help some endangered species.

On a final and totally bizarre note, scientists at Indiana University have matched eggs and sperm from two species of sea urchin, *Heliocidaris tuberculata* and *Heliocidaris erythrogramma*, which are believed to have split from the same ancestor species some ten million years ago. The resulting hybrid larvae show characteristics expressed by neither parent species. Vaguely resembling sea stars, they may be expressing latent genes from their long-extinct parent species. In other words, an animal which lived before mammoths or thylacines ever existed may, in some form, have been brought back. It's not Jurassic Park, but it's certainly food for thought.

THE THYLACINE—DEAD OR ALIVE?

As noted in the last essay, a few researchers believe it's possible to clone the world's largest carnivorous marsupial—the thylacine. The species *Thylacinus cynocephalus*, which resembled a slender, striped wolf, officially became extinct in 1936, when the last known individual died in a zoo. Since then, however, sighting reports and occasional tracks from Tasmania and from mainland Australia have kept alive the slim but tantalizing possibility the animal still exists. Accordingly, some zoologists argue, cloning won't be necessary if we look hard enough.

Hans Naarding, a veteran Tasmanian park ranger, claimed a close-range observation of a thylacine in 1982. He spotlighted the animal at night from a distance of about twenty feet. Nick Mooney, the game official who led a subsequent search, collected enough "excellent" reports to conclude the area "was used by thylacines." This locale was near the headwaters of the Salmon River in northwest Tasmania.

In 1995, there was another sighting by a park ranger. The witness was Charlie Beasley, who spotted a lone animal in northeastern Tasmania. He had a clear view for two minutes and described the animal as sandy-colored and about half the size of a German Shepherd dog. Professor Struan Sutherland quoted the story in a magazine article in which he also described a 1945 expedition led by zoologist David Fleay. This team collected hair, droppings, and footprints of thylacines which had survived the species' presumed demise. (Fleay had another distinction: while photographing the last captive specimen in 1936, he became the last human—we don't know if he's the only one, but certainly the last—to be bitten on the rear end by a thylacine.)

In 1996, an article in *Scientific American* reported the Tasmanian National Parks and Wildlife Service continued to receive "dozens" of sighting reports every year. The continuing flow of sighting and the dearth of accompanying hard evidence prompted author Daniel Drollette to call the creature "Tasmania's version of the Loch Ness monster." One park service official, Mark Holdsworth, lamented that the tiger search was draining resources which could protect known endangered species.

Living thylacines are occasionally reported from the Australian mainland. In 1997, for example, two forestry workers claimed such an encounter. The news report gave the location as "near West Takone." The animal, a "sandy-gray, black-striped, doglike" beast, wandered right out onto the road in front of the witnesses' vehicle. About the same time, there were sighting reports from Irian Jaya, the western portion of the island of New Guinea, of animals resembling thylacines. The WWF was intrigued enough to investigate, and a local government reportedly posted a sizable reward for a live thylacine. To date, however, no one has brought in a thylacine, alive or otherwise. There is no known fossil record of thylacines in this area, although there are fossils from neighboring Papua New Guinea.

Also in 1997, a group called the Australian Rare Fauna Research Association began a new effort to document the continued existence of the thylacine on the mainland of Australia. The Association, founded by the late cryptozoologist Peter Chapple, created a project called "Tiger 2000" to collect evidence for the animal. The Association collected records of well over a thousand sightings, but the supporting evidence was disappointing: one pawprint cast and a photograph Chapple admitted was too blurry to be definitive.

In a 1998 book, Dr. Eric Guiler, the world's leading thylacine expert and a veteran of a decades-long search for the animal, reluctantly admitted he was now among those "who no longer believe in the survival of the Tasmanian tiger."

Still, the current fact sheet from Tasmania's Parks and Wildlife Service notes, "...the incidence of sightings introduces a reluctance among some authorities to make emphatic statements on the status of the species." Translated from the bureaucratese, this may be read as, "There's still hope."

THE CAT CAME BACK

Very next day, the cat came back.
Thought he was a goner but the cat came back
'Cause he couldn't stay away...

— AMERICAN FOLK SONG

The Eastern panther (or puma, cougar, or whatever) is proving to be a very difficult species to keep in the "extinct" category. Ecologist Chris Bolgiano, who lives in western Virginia, has written, "Sometimes it seems I am the only person I know who hasn't seen a panther."

Puma concolor (or *Felis concolor*, for those who believe an older approach to its classification is correct) was once the most widespread predator in the Americas. The Cherokees of the southern Appalachians called the animal *Klandagi*—lord of the forest. The white settlers called it a dangerous predator requiring extermination. After a massive campaign of government-subsidized hunting, the Eastern subspecies in the United States was wiped out by the late 1930s.

Someone may have forgotten to tell the cat.

The cougar is a remarkable animal. It can leap at least thirty feet, jump into a tree while carrying a deer carcass, and generally adapt to all kinds of environments and situations. It is so stealthy that many people live all their lives in cougar country without seeing one, and the only practical method of hunting this species is to use specially-trained dogs. Figuring out just what the cougar is from a genetic and evolutionary standpoint has proven to be a problem. DNA suggests affinities with such disparate animals as the cheetah and the bobcat, but the cougar is so different from all other cats that the Society of Mammologists in 1993 approved placing it in its own genus, Puma.

The Eastern cougar is technically known as *Puma concolor cougar*. From 1995 to 1998, there were over 400 sighting reports of this animal, or something like it, made to the private Eastern Cougar Research Center. More importantly, the first glimmers of supporting evidence have surfaced. Bolgiano, who is very cautious about accepting any sighting reports, nevertheless writes, "I myself have seen a home video filmed in western Maryland in 1992 that showed an unmistakable cougar stepping momentarily between trees in a forest."

Whether cougars survive in the eastern United States is not just an academic debate. There are political and economic as well as scientific implications. If it is established that the Eastern cougar—which is on the U.S. Endangered Species List—still exists in the region, the Fish and Wildlife Service (FWS) will be compelled by law to take action to protect it. This will be expensive and could involve limits on some recreational wilderness use, including hunting.

The stealthy cat's presence would also be an interesting test case for cryptozoology. If a large predator really exists in a well-populated region, identified by hundreds of sighting reports but leaving only minimal hard evidence, there are implications for the survival of other extinct-but-reported animals (notably the thylacine) and for some cases of unproven animals as well.

There already is one similar example—the case of the British big cats. For the past few decades, citizens all over the British Isles have been reporting a puzzling variety of non-native felines, including lions and black leopards. The evidence cited, including kills, pawprints, and distant photographs, was never enough to convince government agencies, although some police searches were carried out. Wildlife officials pointed to the lack of any results from these searches and generally dismissed any thought that big cats could be living wild in Britain.

That attitude changed dramatically in August 2000, when a large cat, apparently a black leopard, clawed the face of an eleven-year-old boy in Monmouthshire. A massive search was completely unsuccessful, but no one now doubts the existence of elusive feline predators. Experts speculated that (as some cryptozoologists, like Karl Shuker, have held for a long time) a 1976 law restricting ownership of exotic felines resulted in the release of enough such cats to establish at least one breeding population. This demonstration that large predators could exist for many years, even in a well-known area, without leaving obvious proof of their presence, is an important one, and the analogy to the Eastern cougar is difficult to dismiss.

For many years, FWS officials, like their British counterparts, rejected any notion of the Eastern cougar's survival. This was not unreasonable. Alleged cougar tracks were rare and often ambiguous, and there were no bodies. Like a cat creeping up quietly on its prey, though, pieces of evidence have slowly fallen into place. A cougar of uncertain origin (it showed signs of a domesticated lifestyle) was killed in Tennessee in 1971. Tracks, hair, and droppings found in New Brunswick in 1992 were identified by the province's wildlife authority as belonging to a cougar. A deer definitely killed by a cougar was found in New York in 1993. The FWS confirmed that droppings found in Vermont in 1994 were from a mother cougar and two kittens. A farmer in Virginia was compensated by the government in 1998 after a cougar apparently killed his goats. Also in 1998, a large, long-tailed cat left clear tracks, reportedly up to eight inches wide (extremely large for a cougar), in the dirt of a Rhode Island family's yard.

According to the official FWS fact sheet on the animal, the best evidence comes from the Great Smoky Mountains National Park, and "there were an estimated three to six cougars living in the park in 1975." The FWS produced a Recovery Plan in 1982 to manage the animal if its existence was confirmed. The same fact sheet reports on a five-year survey by the FWS and U.S. Forest Service from Virginia to northern Georgia. This effort failed to confirm any cougar populations.

There is another problem here. Establishing the existence of cougars in the East does not prove they constitute a native population. Some cougars are kept as exotic pets (often illegally), and there's no doubt some have been dumped when they grew from cute kittens into hundred-pound felines capable of eating their "owners." For example, the 1998 Rhode Island specimen came up to a home and rummaged in a trash barrel. To one expert, Bruce Clark of the Roger Williams Park Zoo, that strongly indicated a domestic origin.

In August 2000, a new series of good reports from Windham County in southern Vermont brought this comment from state biologist Forrest Hammond: "We're ready to believe that there are some cougars that have been seen in the state. We're not going to believe that we have a population until we can confirm it. It's going to take some pretty hard work to do that."

Confirming cougars can take a while. In Illinois, the resident cougar population (considered part of the Eastern subspecies) was believed wiped out by 1862. Numerous reports of a surviving population went unconfirmed and unaccepted by wildlife authorities until July 2000, when a specimen was killed by a railroad train in Randolph County. According to Chris Bolgiano, the necropsy on this cat revealed no sign the cougar had ever been domesticated. Michigan's last resident cougar was supposedly shot in 1906. In 2001, DNA analysis of cat feces, along with deer kills and other evidence, convinced private researchers the occasional sightings since that date had concerned a relict population that was never wiped out.

The genetic difference between the Eastern cougar and other subspecies is slight or nonexistent, so it's difficult to establish a particular cat's origin even if the whole animal is in hand. It's impossible to tell from scat or pawprints.

Then there's the conundrum posed by the "black panther." About a quarter of all the Eastern panther reports involve black animals. No one has ever produced a body or a photograph of an unquestioned black panther on the entire North American continent. Only one or two kills of such animals have been documented in Latin America. Yet the reports keep coming. My father, Don Bille, an experienced hunter, saw a black feline cross the road in front of his car in Maine in the 1950s. He described it as much too large for a feral domestic cat, although smaller than black leopards he'd seen in circuses. The IUCN Cat Specialists Group Species Account says only, "Melanism has been widely reported."

Since the Eastern cougar population has been very small for a long time, it's been suggested it would only take one cat carrying the gene for melanism to spread this trait among a considerable fraction of the animals remaining today. Another possibility is that there are only ordinary-colored panthers, and the witnesses reporting black ones are mistaken. A third theory is that one or more black leopards have escaped captivity and have lived long enough to be reported as black panthers, but no specific instance of such an escape or release has been documented.

Whatever the solution to the color question, it now appears very probable there are panthers roaming the eastern U.S. It has still not been proven they were there all along, although it seems likely that they have. The sightings never stopped—for example, a biologist spotted a cougar in his headlights in New Jersey 1958—and it's hard to argue that the millions of acres of protected forest along the Appalachians couldn't have sheltered a small breeding population. Cougars may have ranges exceeding a hundred square miles per cat, so the animals would be spread out and safe from all but chance detection.

I'll admit to a bias here. I want there to be Eastern cougars. I want to think humanity failed in an all-out effort to wipe out this cat, and I want to know a viable population of large mammals can stay hidden for decades. Am I correct? As is so often the case, only time will tell.

THE HOWLING GOD

The wolf has captured humanity's imagination in a way no other creature ever has. Unfortunately, such notoriety has not been good for this intelligent and reclusive predator.

In North America, a relentless campaign of extermination has resulted in the demise of at least six subspecies of *Canis lupus*. Not surprisingly, two of the most spectacular types were among the first to depart this planet. The Newfoundland wolf, weighing well over 100 pounds, became extinct about 1911. The even larger Kenai wolf probably vanished from its Alaskan range about 1915.

The smaller red wolf, *Canis rufus*, almost disappeared as well. Weighing 40 to 70 pounds and clad in a tawny and cinnamon coat with black or gray markings, this species was restricted to the southeastern U.S. The Texas subspecies is extinct after having survived into the early 1970s. The Mississippi red wolf, driven into the most miserable swamps of Texas and Louisiana, was extinct in the wild by 1980. Its genetic shadow lives in a population of hybrid coyotes into which the last handful of wild red wolves—inbred, sick, and unable to find mates of their own kind—deposited their legacy. (Some authorities believe the red wolf always was a "hybrid swarm" of gray wolf and coyote genes, or that the original "pure" red wolf was a subspecies of gray wolf. American conservation authorities, though, continue to treat it as a species.)

Fortunately, a captive population was bred from fourteen wolves trapped between 1973 and 1979. Beginning in 1987, a few wolves were released onto government-owned land in North Carolina, and other reintroduction projects are underway. There are now over 300 red wolves in captivity or in the reintroduction areas.

Wolves may, at one, time, have been the most wide-ranging non-human mammals on Earth. Even the island nation of Japan had its wolf—the *yamainu*, or *shamainu*. The yamainu is normally listed as a wolf subspecies, *Canis lupus hodophilax*, although the animal was originally described as a separate species, *C. hodophilax*. A few modern specialists, such as Dr. Yoshinori Imaizumi, have argued this unique identity should be resurrected.

If the yamainu was a true wolf, it was the smallest known race of that species. It was only about fourteen inches high at the shoulder and had disproportionately short legs for a wolf of any sort.

Centuries ago, the diminutive wolf was looked on favorably in Japanese culture. Folk tales often cast the animal as a friend or guardian of humans. Despite this, the yamainu still had the air of mystery humans have always ascribed to wolves. This may have been due to the species' vocalizations. These wolves reportedly howled for hours at a time, giving voice to a haunting wail that seemed far too loud to come from such a small animal. To the Ainu, the indigenous people of Japan, the yamainu was the "Howling God."

As Western-style farming and the keeping of domestic animals increased, the yamainu took on more of the aura of a threat. By the late nineteenth century, the yamainu had, in many areas, reverted to the role of wolves everywhere—that of enemy of the human race. It was hunted for its fur, because of the alleged threat it posed to livestock, and because of the danger posed by occasional rabid specimens. In areas where wolf sightings were once ignored or welcomed, villagers posted magical charms to ward off wolves or actively pursued the animals with traps, weapons, and poison. Governments placed bounties on the animals. As the species became rarer (and thus even less of a real threat), the persecution intensified—an illogical reaction which has occurred many times in many lands where wolves are concerned.

It is commonly believed that the last Japanese wolf was killed in 1905. At the site of this event, in Higashi-Yoshino Village in Nara Prefecture (one of the southernmost prefectures, or states, on the main island of Honshu), a requiem ceremony is carried out each year. The Japanese have exterminated their wolf, but not without regret.

Or have they?

Occasional sighting reports have kept the question of the Japanese wolf's existence open for almost a century now. In 1934, for example, a group of farmers northwest of Hongu reported seeing five or six wolves in a pack. After World War II, sightings increased. According to forester and writer Ue Toshikatsu, this was a logical development, since conscription and war reduced the population of rural areas and produced an increase in the numbers of wild game such as boar and deer.

Sighting reports dropped off again in the late 1950s. Ue noted that, in this period, forests and wildlife were again under pressure from development. He suggested the wolf survived its presumed demise and began a modest comeback during and after the war, only to meet a final end around 1960.

In 1993, Yanai Kenji published his own story of how, as a mountaineer accompanied by his son and his co-worker, he was startled by a "horrible howling" while hiking toward Ryogami Mountain in 1964. Soon after hearing the howls, the party encountered a lone wolf. The animal watched them briefly, then fled, leaving the half-eaten carcass of a hare behind.

In March 1994, wolf enthusiasts hosted a conference in Nara. Over eighty professional and amateur researchers attended. They presented and analyzed reports from seventy witnesses who had seen wolves or heard howls. An accompanying story in the *Nihon Keizai Shimbun* stated that a shrine in Tottori Prefecture, just northwest of Nara, was discovered in January 1994 to hold a surprisingly recent specimen of the yamainu. This animal may have been presented to the shrine as recently as 1950.

The largest number of modern sightings have come from the Kii Peninsula. This rugged, mountainous block of land projecting into the Pacific from the southeastern coast of Honshu was the last stronghold of the yamainu.

The photographic evidence, though, has come from other regions. In 1966, a wolf enthusiast named Hiroshi Yagi was driving on a forest road in the central Japanese prefecture of Saitama (well north of the Kii) when he spotted what he believed was a wolf. He stopped, and the animal let him get close while he took photographs. These pictures sparked a renewal of interest, though not all authorities accepted they showed a wolf rather than a domestic dog.

The next claim came from the southern island of Kyushu. On July 8, 2000, school principal and amateur biologist Akira Nishida was hiking on a mountain trail when a canid about three feet in length emerged from the forest and passed within a few yards of him. Nishida whipped out his camera and shot ten pictures. The aforementioned Dr. Imaizumi examined the evidence and said, "I cannot help but think that the animal is a Japanese wolf." Other scientists were cautious about identifying the animal, but agreed the case deserved investigation. A claim that this, too, was an abandoned dog has divided expert opinion.

Recent expeditions focused on the Kii Peninsula have produced no new evidence, despite trapping efforts and the playing of recorded Canadian wolf howls. Modern sighting reports that don't include photographs are hard to interpret: the wolf is beyond living memory in Japan, and some witnesses, at least, no doubt are misidentifying dogs.

Folklorist John Knight suggests the wolf has become a symbol, or metonym, for the place the mountain forests once held in Japanese culture. Such forests were viewed as lands of mystery and danger. In short, the continuing interest in wolves exists because people want to believe the wildness, the mystery, is still "out there," despite the slender nature of the evidence. This may well be true, but it doesn't answer the question: is the wolf extinct, or are there nights when, on Japan's loneliest mountaintops, the Howling God still speaks?

SECTION III: THE CLASSIC MYSTERY ANIMALS

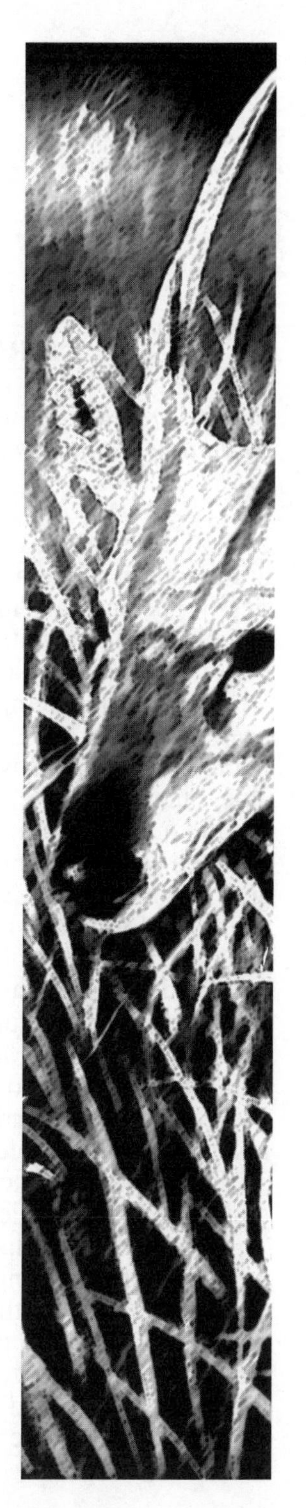

INTRODUCTION: SECTION III

Reality provides us with facts so romantic that imagination itself could add nothing to them.

— JULES VERNE

It should be clear by now there are still many undescribed species hiding out in the remote areas of the world. But what about the stories of spectacular "monsters?" Could there be yetis, sasquatches, and plesiosaurs awaiting discovery, or are such things just the product of the human imagination?

This section attempts a balanced look at this question. There has been so much hype, misinformation, and exaggeration involved that most zoologists wouldn't touch the subject. This is understandable, but unfortunate. It may be that none of these creatures are real, but the questions surrounding their existence are worth examining.

Even for the skeptic, there are nagging bits of evidence which are hard to dismiss. Two examples are the still-puzzling yeti footprints photographed by Eric Shipton in the Himalayas in 1951 and the description of a large, long-necked marine beast recorded by two British naturalists in 1905.

Since we are venturing here into the most controversial aspect of a controversial field, it's appropriate to examine concerns about the validity of cryptozoology as a branch of science. Critics, such as Professor Robert Carroll (author of *The Skeptic's Dictionary*), often assume 1) that monster-hunting is the whole of cryptozoology, and 2) that such an endeavor should be lumped with UFOs, hauntings, and other "pseudosciences."

Such a flat-out dismissal of cryptozoology is unjustified. The criterion for admitting a given field of inquiry to the ranks of the sciences is whether that field deals in testable hypotheses—that is, ideas which can be tested and proven to be true or false. Given enough resources, every proposition made in cryptozoology can be empirically validated or invalidated.

For example, take the hypothesis, "There is an unclassified ape living in the Pacific Northwest of North America." A sufficiently large and thorough search operation would either find the animal or establish its non-existence. The resources needed for such a definitive search have not been available, but that is not important here. The point is that cryptozoology—even the "monster" portion of the field—deals with the question of whether particular animals exist in particular environments. One may argue that no such animals exist, but there are no substantive arguments for the proposition that cryptozoology is *a priori* unscientific.

Critics often point out that some cryptozoologists do unscientific things, such as accepting weak evidence without proper scrutiny. While this is (regrettably) true, it's not relevant to the question of whether cryptozoology is a proper science. When Stanley Pons and Martin Fleischmann enthusiastically announced the discovery of cold fusion, only to have their claims disproved, the results showed their methods of investigation were flawed. No one argued this affair meant nuclear physics was invalid as a science. To reiterate: because cryptozoology consists of the investigation of testable hypotheses, it is legitimately scientific in principle, if not always in practice.

Saint Columba confronts the Loch Ness Monster, A.D. 565.

The hypotheses being examined in this section include the existence of the yeti, sasquatch, the "sea serpent," and several famous "lake monsters." Perhaps these "cryptids" (as unconfirmed animals are sometimes called) are all mythical, although I'll be surprised if we don't eventually turn up at least one new species of large marine animal. If they are real, then science will be far richer because a few oft-ridiculed enthusiasts kept up the hunt. If they aren't, we will still gain knowledge from the chase—knowledge of the animal world, and of our own reasons for believing there are still monsters to be discovered.

THE BRITISH NATURALISTS' SEA MONSTER

Are there large and strange unclassified animals roaming the oceans of the world? The best single piece of evidence to date on this question came from two British men of science, Michael J. Nicoll and E.G.B. Meade-Waldo. In 1905, these witnesses observed a "sea monster" which has never been explained.

The men were both experienced naturalists, Fellows of the Zoological Society of London. Their account of "a creature of most extraordinary form and proportions" is recorded in the Society's *Proceedings* and Nicoll's 1908 book *Three Voyages of A Naturalist*.

On December 7, 1905, at 10:15 AM, Nicoll and Meade-Waldo were on a research cruise aboard the yacht *Valhalla*. They were fifteen miles east of the mouth of Brazil's Parahiba River when Nicoll turned to his companion and asked, "Is that the fin of a great fish?"

The fin was cruising past them about a hundred yards away. Meade-Waldo described it as "dark seaweed-brown, somewhat crinkled at the edge." The visible part was roughly rectangular, about six feet long and two feet high.

As Meade-Waldo watched through "powerful" binoculars, a head on a long neck rose in front of the frill. He described the neck as "about the thickness of a slight man's body, and from seven to eight feet was out of the water; head and neck were all about the same thickness ... The head had a very turtle-like appearance, as also the eye. It moved its head and neck from side to side in a peculiar manner: the color of the head and neck was dark brown above, and whitish below—almost white, I think."

Nicoll noted, "Below the water we could indistinctly see a very large brownish-black patch, but could not make out the shape of the creature." They kept the creature in sight for several minutes before the *Valhalla* drew away from the beast. The yacht was traveling under sail and could not come about. At 2:00 AM on December 8th, however, three crewmembers saw what appeared to be the same animal, almost entirely submerged.

In a letter to author Rupert T. Gould, author of *The Case for the Sea Serpent*, Meade-Waldo remarked, "I shall never forget poor Nicoll's face of amazement when we looked at each other after we had passed out of sight of it ... " Nicoll marveled, "This creature was an example, I consider, of what has been so often reported, for want of a better name, as the 'great sea-serpent.'"

What did these gentlemen see? Meade-Waldo offered no theory. Nicoll, while admitting it is "impossible to be certain," suggested they had seen an unknown species of mammal, adding, "the general appearance of the creature, especially the soft, almost rubber-like fin, gave one this impression." The witnesses did not notice any diagnostic features such as hair, pectoral fins, gills, or nostrils.

The late zoologist Bernard Heuvelmans, in his exhaustive tome *In the Wake of the Sea-Serpents*, suggested this sighting involved a huge eel or eel-shaped fish swimming with its head and forebody out of the water. For reasons no one understands, the largest known

The "sea serpent" reported by British naturalists Nicoll and Meade-Waldo in 1905.

species of eel, the conger, does swim this way on occasion. Interestingly, the conger also has been observed to undulate on its side at the water's surface, producing an appearance that looks little like an eel and a lot like a serpentine monster, albeit a small one. Congers are known to reach about nine feet in length.

Another candidate for the sighting might be a reptile. Nicoll's sketch certainly bears some resemblance to a plesiosaur, a Mesozoic-era tetrapod suggested as a solution for sea serpent sightings as early as 1833.

Plesiosaurs keep turning up in connection to sea serpents because they were one of the few marine species of any type in the fossil record to have long necks. American humorist Will Cuppy once remarked on plesiosaurs, "They might have a had a useful career as sea serpents, but they were before their time. There was nobody to scare except fish, and that was hardly worth while." Indeed, the plesiosaur fossil record stops with that of their land-based cousins, the dinosaurs.

There is another problem in connecting these animals to the 1905 description. In addition to the absence of relevant fossils dated within the last sixty million years, no plesiosaur is known to have possessed a dorsal fin. There was no need for a dorsal fin for stability on the turtlelike bodies of these animals. A plesiosaur with a fin or frill unsupported by bones and thus unlikely to fossilize, presumably for threat or sexual display, is not impossible, but this is pure speculation.

Nicoll's idea of a mammal poses problems as well. No known mammal, living or extinct, fits the description given by the two naturalists. Some cryptozoologists believe sea monster reports are attributable to archaeocetes: prehistoric snakelike whales, such as those in the genus Basilosaurus. It's conceivable this group could have evolved a long-necked form, but the known whales were actually evolving in the opposite direction, resulting in the neckless or almost neckless modern cetaceans. One other mammalian possibility is a huge elongated seal. This seems equally difficult to support, given that no known seal, living or extinct, has either a truly long neck or a dorsal fin.

Meade-Waldo was aware of the famous sea monster report made in 1848 by the crew of the frigate H.M.S. *Daedalus*. He thought his own creature "might easily be the same." The *Daedalus* witnesses described an animal resembling "a large snake or eel" with a visible length estimated at sixty feet.

There are a few reports specifically describing giant eels. A German vessel, the *Kaiserin Augusta Victoria*, observed such a creature in its entirety off England in 1912. The *Kaiserin's* Captain

Ruser described it as about twenty feet long and eighteen inches thick. Four Irish fisherman claimed to have caught a nineteen-foot eel in 1915. In 1947, the officers of the Grace liner *Santa Clara* reported their ship ran over a brown eel-like creature estimated at sixty feet long. In 1971, English fisherman Stephen Smith was in the area of the 1912 sighting when he allegedly encountered an eel over twenty feet long, with the head of a conger eel but "four times the size." He told author Paul Harrison, "I have fished all over the world, but never have I seen something like this." Smith suggested it was… "a form of hybrid eel, but at twenty feet? There must be a more rational explanation, but I'm damned if I know what it is!"

The only "non-monster" hypothesis which has been advanced to explain the *Valhalla* sighting came from Richard Ellis, a prominent writer on marine life. Ellis has suggested that a giant squid swimming with its tentacles foremost, with one tentacle or arm held above the surface, could present an unusual appearance which, combined with a reasonable degree of observer error, might account for the details reported in this case.

Squid can swim tentacles-first, and often do so when approaching prey. For one to have presented the appearance described, though, it must have acted in a totally unnatural fashion. The squid would have to swim on its side to keep one fin above the water while pointlessly holding up a single limb and swimming forward for several minutes. Even assuming it is physically possible for a squid to act this way, it seems impossible to come up with a reason why it might do so. This explanation also requires that Meade-Waldo, at least, made a major mistake, since he recorded seeing a large body under water "behind the frill."

While the idea of a large seagoing animal remaining unidentified to this day may seem surprising, it's not beyond the bounds of plausibility. Recently identified whales have already been mentioned. The sixteen-foot megamouth shark (*Megachasma pelagios*) was only discovered when caught by accident in 1976. A unique feature of the megamouth case is that this species—a slow-moving, blimplike filter-feeder which became the sole inhabitant of a new family—was not just unknown as a living species, but completely unknown in every respect. There were no fossil indications, no sighting reports, and no local folklore about such a strange creature among Pacific islanders. The species just appeared. Finally, as we will see in Section IV, at least one type of whale is generally accepted by cetologists despite a lack of physical evidence.

We are left with this simple fact: on December 7th, 1905, two well-qualified witnesses described a large unknown marine animal for which no satisfactory explanation has been presented. Their report strongly indicates the oceans hold (or held at that time) at least one spectacular creature still evading the probes of science.

THE PRIMATE PROBLEM

Of all the mammals, it is our relatives, the primates, which attract the most popular interest. This is understandable, as is the fascination with the idea that more primates, including large ones related to humans, may still await discovery.

The idea of unknown humanlike primates has been floating around our collective consciousness since humanity's time began, but a thunderbolt of a discovery from the Indonesian island of Flores, announced October 27, 2004, raised the concept to a new level of awareness. When scientists led by Australians Mike Morwood and Peter Brown and Indonesian Thomas Sutikna found the three-foot human relative, *Homo floresiensis*, cryptozoologists were almost as excited as anthropologists. *H. floresiensis* might have lived as little as 12,000 years ago, when it would have coexisted with the larger modern humans. Paleontologist Henry Gee of the journal *Nature*, which published the report, commented, "They are almost certainly extinct, but it is possible that there are creatures like this around today." Dr. Gee added, "Large mammals are still being found. I don't think the likelihood of finding a new species of human alive is any less than finding a new species of antelope, and that has happened."

The natives of Flores had tales of the *ebu gogo*, little men about three feet tall who, the islanders say, was still around only a century ago. From 12,000 years to 100 is a big leap, but it's hard to call it an impossible one. Some scientists, inevitably, have begun to refer to the dwarf humans of Flores as "Hobbits." (A humorous note is that, just before publishing the discovery of Flores man, Dr. Gee had completed a book called *The Science of Middle-Earth*.)

With the shock waves of the Hobbit discovery still reverberating through the world of anthropology, let us turn to the claims that unclassified living primates—some smaller than man, some the same size, some considerably larger—still haunt the Earth.

Such creatures are reported from every inhabited continent except Europe (and even there, are represented in old folklore about "wild men"). This global distribution is one of the major stumbling blocks in obtaining scientific credibility for the present topic. It is one thing to ask a primatologist to accept there is an unclassified ape at large. It is quite another to suggest the planet houses several good-sized bipedal primates, all uncaught and unclassified. Such a claim, made seriously by some cryptozoologists, is extremely difficult to even consider without hard evidence. That, in turn, makes many specialists unwilling to admit that any of these alleged animals could exist, the Flores discovery notwithstanding.

This problem raises the standard by which evidence is judged. The current situation can be stated very simply: the only evidence likely to result in widespread acceptance of any unknown large primate is a type specimen.

As of now, no case meets that standard. The yeti is known mainly from tracks and local traditions, plus a few reports by Westerners. The evidence for other primates, including Africa's *dodi* and *kikomba*, Australia's *yowie*, China's *yeren*, Siberia's *chuchunaa*, and South America's *di-di* and *sisemite*, is similar or weaker. North America's sasquatch adds some disputed film evidence and many recent sightings. (The yeti and sasquatch, as the most famous primates in the crypto-zoo, merit more detailed essays later in this section.)

The dodi, one of Africa's alleged mystery apes.

Anthropologist Myra Shackley has put forth a great deal of effort to prove relict Neanderthals, known as *almas* and by many other names, inhabit Mongolia, the Pamirs, and the Caucasus mountains. Unfortunately, she, too, has turned up no evidence more concrete than footprints and anecdotes. Along the rugged border shared by Pakistan and Afghanistan—a region in the news lately due to the search for terrorist Osama bin Laden—the local "wild men" are known as *barmanu*. In August 2002, Spanish zoologist Jordi Magraner was searching for these creatures when he was murdered. His death was a sobering reminder that finding new animals in remote regions is neither simple nor safe. (Reportedly, some local men involved in the region's chronic border conflicts assumed Magraner's communications gear marked him as a spy.)

Cryptozoologists must be open-minded, and generally hate to write off a seemingly insubstantial story which may prove to have been important. Still, some prioritization is necessary. Researchers must sort through the many reported primates and decide which cases, if any, most plausibly point to a real animal.

Sasquatch or bigfoot is the most widely reported of all such creatures, but is still highly problematical. There are thousands of sightings and footprint reports, and supporters say it would it be impossible to fake them all. As Craig Woolheater of the Texas Bigfoot Research Center put it, "What it comes down to is that if just one (witness) is telling the God's honest truth ... then there's something out there."

Critics, like anthropologist Kevin Wylie, respond that the problem arises when you examine these reports in detail. To Wylie, none of the sightings or other pieces of evidence is impressive enough to encourage acceptance of something as improbable as a huge, undiscovered North American ape.

Sasquatchlike creatures are widespread in Native American lore, although the origins and meaning of these stories are difficult to evaluate. The Salish word from which "sasquatch" is derived refers to a supernatural creature, not an animal. On the other hand, zoologist Ivan Sanderson wrote in the 1960s that, when one Indian was asked about the subject, the reply was a derisive, "Oh, don't tell me the white men have finally gotten around *to that*."

The late Dr. John Napier was the most prominent primatologist ever to examine this problem in depth. In his 1972 book *Bigfoot*, Napier endorsed sasquatch as a real animal, although he doubted all other such primates.

Concerning the almas, it seemed to Napier the inhabitants of its rugged homeland make the animal sound too plentiful, as if it should be easy to find. Napier recounted the story of one Caucasus resident who was asked if the almas was mythical. The man, proud of his people's rich mythology, was actually offended that anyone would think it included something as common and boring as the almas. It was reported in 1985 that 5,000 almas sightings and fifty footprint reports were on file with the U.S.S.R. Geographical Society.

Some cryptozoologists suggest it is this creature which is represented by Bernard Heuvelmans' *Homo pongoides*, the apelike man. Dr. Heuvelmans did publish a description and identify an alleged type specimen. This was the famous Minnesota Iceman, a hairy corpse six feet tall. Heuvelmans, along with Ivan Sanderson, examined this traveling exhibit (frozen in a block of ice) in 1968. Heuvelmans was convinced the thing was genuine, although his colleague had reservations. The Iceman was being shown by one Frank Hansen, who told at least three different stories of its origin. Heuvelmans believed the Iceman was actually shot in Vietnam during the U.S.-Vietnamese war and then smuggled into this country. However, he had only secondhand reports to substantiate these events. There are ongoing reports of apelike primates in Vietnam (the *Ngoi rung*, or forest people), but no evidence connecting them to the Iceman.

Interviewed in 2000, Hansen said he never did know what the thing was. "It's history," he told a reporter. "I don't care to get involved in that." Such a comment, while not constituting a confession, is unlikely to reinforce the Iceman's credibility. That credibility was already in question thanks to a 1981 newspaper article which claimed the Iceman was a creation of the late Howard Ball, a maker of animal models for Disney. Ball's widow and son supported the story.

Even if the Iceman was not—as seems likely—a clever hoax, the thing's current whereabouts are unknown. It isn't unknown for science to accept a description based on missing evidence. After all, no one doubts the validity of Peking Man, whose bones have been lost for over sixty years. *H. pongoides*, however, represents such a startling claim that the reluctance to accept it is understandable unless the mysterious corpse surfaces again.

Just to further confuse the issue: as noted above, many who accept the Asian ape-man reports believe they concern Neanderthals. Heuvelmans was one of these, and felt his *H. pongoides* was an example of such a survivor. However, Sanderson, the only other scientist to see the Iceman, strongly dissented. Not only did he feel the Iceman might have been an expertly-made model, but he wrote, "This creature is almost as far removed from the standard neanderthaloid construction as is possible." Heuvelmans' supposition was based on the controversial belief that the common reconstructions of Neanderthals as broad-faced, heavy-browed people are substantially inaccurate.

The yeti is in a state similar to that of the almas. Despite the long-standing local traditions and the efforts of determined cryptozoologists, the beast remains elusive. After a half-century of serious investigation, the best yeti evidence is still the broad, strange-looking footprints photographed by mountaineer Eric Shipton in 1951. Reported yeti remains have either been lost or identified as having belonged to known animals.

So the alleged large primates of the world are supported, at best, by an interesting but inadequate collection of local traditions, reports, footprints, and hair samples. For now, the motto of science remains *habeas corpus*, or "bring the body forth!"

IS THE YETI STILL OUT THERE?

Forty years ago, the world's pre-eminent mystery animal was the yeti (also known as *meh-teh*, "abominable snowman," etc.). Prominent men like Sir Edmund Hillary went in pursuit of it, and well-qualified zoologists speculated openly about an unknown species of great ape.

Today, nearly everyone seems to have lost interest. Why?

Researchers who still seek large unknown primates have mostly shifted their focus to the sasquatch, and, to a lesser extent, the yeti's neighbors, central Asia's *almas* and China's *yeren*. This may be because the sasquatch evidence is more recent and seemingly more solid. While there is still no carcass, there are alleged films and photographs, more sightings, more footprints, and so on.

The chief trouble with sasquatch is that it lives in a populated (although still largely undeveloped) area. Skeptics have a hard time accepting there's never been a specimen accidentally killed by a hunter or hit by a car. With the yeti, the problem is the opposite. There are high, forested valleys in the Himalayas almost untrodden by humans, but there is very little in the way of evidence to show anything like the yeti lives in those valleys. Yeti "scalps" have been proven to be imitations, and preserved "hands" to be from known animals (one possible exception is discussed below). While there is a rich body of local beliefs and stories concerning these animals, under various names like *kang-mi* ("snow-man"), such evidence is difficult for Western scientists to evaluate. Some traditions specify there are two or three species of yeti, from one smaller than a human being up to a giant, eight feet tall, which sounds much like the North American sasquatch.

Concerning sightings by Westerners, there are only a few whose authenticity is not disputed. All involved creatures seen for a short time at a considerable distance.

In 1925, nature photographer N. A. Tombazi had a "fleeting" glimpse of a humanlike figure. Don Whillans saw an apelike animal bounding across the snow at night in 1970, and British mountaineer Chris Bonington reported a similar encounter the same year. The Italian Reinhold Messner, a legend among climbers as the first man to scale Mount Everest without oxygen, reported a close meeting in 1986, but that sighting has become enmeshed in a controversy discussed later in this essay.

The only physical evidence in favor of the yeti is a desiccated hand formerly kept at Pangboche monastery. The relic was reported stolen in 1991, and the monastery burned down a year later. However, yeti-hunter Peter Byrne claims to have stolen the thumb and phalanx of this hand in 1959. (He has since expressed his regret for the theft.)

Tests on the samples provided by Byrne allegedly identified them as primate but not human. These results were reported on NBC's television series *Unsolved Mysteries*, but have never been published for scientific review. Photographs show a humanlike hand, too broad to have come from one of langur monkeys which inhabit the Himalayan region. No one knows the age of the hand. Ivan Sanderson suggested it was human but ancient, and was possibly the long-preserved hand of a Neanderthal.

The best evidence in the yeti file is a 1951 series of photographs by mountaineer Eric Shipton from a glacier in the Menlung Basin at 19,000 feet in the Himalayas. Shipton said he and his companion, Michael Ward, found huge tracks in the snow and followed them for about a mile. Shipton later wrote that he had seen similar tracks before, but this is the first time he took photographs. He thought the tracks were fresh, probably less than twenty-four hours old.

The close-up photographs of one track, with an ice ax and a climbing boot positioned for scale, show a footprint about thirteen inches long and eight inches wide. The print appears anthropoid, although its breadth, the long second toe (longer than the great toe), and the separation between the first two toes and the three smallest ones distinguish it from the prints of known primates.

John Napier wrote that he would dismiss the yeti except, "The Shipton print ... is the one item in the whole improbable saga that sticks in my throat." Noting the print's heel appeared to have melted and refrozen somewhat, Napier suggested the print was not as fresh as Shipton thought. He theorized the print "was composite, made by a naked human foot treading in the track of a foot wearing a leather moccasin," with the sun melting the two together.

There are two problems with this theory. First, we are not dealing with "the Shipton footprint." Shipton and Ward reported a trail of identical tracks a mile or more long. Second, this book's author, at the risk of being thought an idiot by his neighbors, has tested Napier's theory and found it wanting.

In an attempt to reproduce the 1951 tracks, I have made tracks in snow at varying angles to the sun with several types of footwear and then tread on them barefoot. Even though I repeated this experiment several times, trying it with different textures of snow and under varying weather conditions, it simply doesn't work. The toes never lengthen while the whole print widens, which is what would have to happen to get a Shipton-type print. The toes melt together into a blob. The print does expand all around, but never gets wide enough to resemble Shipton's track. Theories suggesting the track was made by melting together of all four pawprints of a small animal, or melted-out pawprints of a bear, meet the same objections. There is no chance of getting a trail of clear, identical footprints from such circumstances.

Conservationist Daniel Taylor-Ide, in his book on the yeti, lamented that if Shipton had photographed several prints, not just one, the animal's stride could be examined and a great deal more information could be deduced. (Many books include an alleged picture of the yeti's trail, but this actually shows a goat's trail. This error was apparently misunderstood by skeptical researcher Joe Nickell, whose 1995 book *Entities* claimed that Shipton photographed nothing but goat footprints.)

Despite the validity of Taylor-Ide's point, the footprint which was photographed begs explanation. It has been suggested that Shipton and his companion, Michael Ward, hoaxed the print, but there is no evidence to prove it, and neither man (both are now deceased) ever admitted to such a trick. Ward has suggested the prints could have been left by a man (some ascetics in the region do walk barefoot in snow) with deformed feet. Given the size of the prints, though, this seems unlikely.

Zoologist Wladimir Tschernezky published an analysis of the footprint (in the prestigious journal *Nature*, no less!) in 1960. He believed a plaster reconstruction of the yeti's foot resembled a gorilla's foot more than a human's and suggested its features were "probably characteristic of early prehominids." He concluded the footprint belonged to a "huge, heavily built bipedal primate, most probably of a similar type to the fossil Gigantopithecus."

One cannot poke into the yeti business for very long without hearing the name Gigantopithecus. This presumably long-extinct animal is also brought up by believers in sasquatch. The reason is simple: this is the only primate known from the fossil record which is large enough to fit all yeti and sasquatch descriptions. Actually, most yeti reports could actually fit an animal considerably smaller than the main species, *Gigantopithecus blacki*, which weighed an estimated 900-1000 pounds. *Gigantopithecus gigantea*, dated to about 6.3 million years B.P., (five million years before *G. blacki*) was only about half that size, although it apparently did not survive in this form but evolved into the larger species.

"Giganto" is usually depicted as an oversized gorilla, although its closest living relation is actually the orangutan. It presumably had a knuckle-walking gait like the gorilla. That is not a definite fact, since the only remains of this primate are three jawbones and hundreds of teeth. The knuckle-walking posture is assumed in most reconstructions and is accepted by paleoanthropologist Russell Ciochon and his co-authors in *Other Origins*, the only authoritative book on Gigantopithecus.

At least one anthropologist, the late Dr. Grover Krantz of Washington State University, dissented. Krantz argued the wide angle at which the bones of the fossil jaw diverge indicated the animal was bipedal, as some of the earlier authorities on the species had suggested. Professor Ciochon responded that every known ground-dwelling ape is basically a quadruped, and there is as yet no evidence that Giganto was so unlike all his relatives. He believed Krantz inferred far too much from the fossil jaws when he claimed they indicate erect posture. Furthermore, Giganto was so heavy that it seems much more likely to have evolved as a quadruped.

As noted, Tschernezky and others have suggested the Shipton footprint looks something like a gorilla's. However, this comparison was made using a mounted gorilla foot, which is not representative of the animal in life. Unfortunately, there are no good tracks identical to Shipton's. Zoologist Edward Cronin, in 1972, reported smaller tracks he thought might belong to a juvenile creature of the type indicated by the Shipton prints, although the resemblance is debatable. There are other reports of footprints looking somewhat like Shipton's, but no good photographs of anything that clearly belongs to the same species (assuming there is a species, of course).

Aside from an episode in 1986, where physicist Tony Woolridge thought he had photographed a yeti (he later agreed he had taken pictures of an unusual boulder), there have been few recent reports of new evidence. A video of unverifiable authenticity, known as the "snow-walker" footage, surfaced in 1996. This was supposedly taken by Belgian hikers in Nepal, but the account was totally lacking in corroborating details (date, location, etc.). Dr. Jeff Meldrum, an American anatomy professor who investigated, found it was likely faked to promote a television documentary.

Not surprisingly, many scientists and researchers, even cryptozoologists, remain skeptical of the yeti. Grover Krantz felt that, as long as the Shipton photographs were the only good non-anecdotal evidence in the yeti file, there was simply no point in looking into the subject. Herwig Zavhorka, a forester who spent three years in the Kaghan Valley in the Himalayas, wrote in 1997 that the yeti was a mix of sightings of the Himalayan langur monkey (*Presbytis entellus*) and the actions of local people who tell visitors what they want to hear.

In 1998, Reinhold Messner published a book he said "solved" the yeti mystery. Messner believed the yeti, or *chemo* as his Tibetan associates called it, was a bear. Messner hypothesized a new species, or a subspecies of the brown bear (*Ursus arctos*), that habitually walks upright, travels by night, and communicates by whistling. Such a bizarre bruin—allegedly standing up to ten feet tall—would be every bit as sensational a discovery as an unknown ape. Messner claimed several yeti sightings and, at one point, said he possessed a yeti skeleton, but he has never produced this evidence for examination. Nor did he discuss this skeleton in his book, which includes pictures of quite ordinary-looking brown bears Messner was told were chemos. Not surprisingly, the cryptozoological community harbors strong doubts about Messner's bear theory.

An interesting bit of yeti news surfaced in April 2001. An expedition in Bhutan had collected a single long black hair stuck to a tree in an area where yetis had been reported. The hair follicle was intact, which allowed for DNA testing. Brought back to England and subjected to tests at the Institute of Molecular Medicine in Oxford, the hair proved a puzzle. Genetics professor Bryan Sykes, a pioneer in using DNA analysis in archaeological studies, told the magazine *New Scientist* he was mystified. "It's not a human, it's not a bear, nor anything else that we've so far been able to identify. We have never encountered any DNA that we couldn't recognize before."

So where does the yeti stand today? As puzzling as they are, strange footprints and an unidentifiable hair are not enough to settle the question. There is, as yet, no good explanation for these items, but they are not sufficient proof to most zoologists that the yeti is a real animal. The yeti case is best marked, "Pending, Awaiting Further Evidence."

BYE-BYE, BIGFOOT?

Three times in the last decade, the sasquatch legend has loudly been proclaimed dead. Whether one thinks sasquatch was ever alive or not, claims to have buried North America's most famous mystery animal once and for all need to be carefully examined.

The first salvo was fired in the September 1998 issue of *BBC Wildlife*. Documentary producer Chris Packham, of the television series *The X-Creatures*, claimed he personally had "killed sasquatch."

Packham's assertion focused on the most famous piece of sasquatch evidence—the 16mm film shot in October, 1967, near Bluff Creek, California, by Roger Patterson. In the article and his subsequent TV program, Packham offered two pieces of evidence: a document he claimed impeached Patterson's story, and a statement by Patterson's companion, Bob Gimlin, that he (Gimlin) might have been the victim of a hoax. This is difficult to verify, since it was rare for Gimlin to talk about the case, and he apparently never made such a statement to anyone else. (In subsequent statements, Gimlin has insisted the original story of the filming was true.)

On closer examination, Packham's argument is very weak. Keep in mind I think it most likely sasquatch does not exist, so I can hardly be accused of blindly defending the animal's validity. Nevertheless, nothing disclosed so far supports Packham's claims to have disproved sasquatch once and for all.

According to Packham, since he has allegedly debunked the Patterson film, "A stake protrudes from the bleeding heart of Bigfoot, from cryptozoology itself." This is an absurd overstatement. Even if he had disproved the film, the film is not the only evidence cited for sasquatch. Moreover, there are plenty of cryptozoologists who do not believe sasquatch exists. Even if the film was a hoax and sasquatch is a myth, that in no way invalidates the entire field.

Packham offered an "incriminating" document he found when visiting Patterson's widow (Patterson died in 1974). This, however, was merely a record of Patterson selling the film rights after the film had been shot. This hardly proves a hoax. Patterson was a man of limited means, often in debt, and it would have been very puzzling if he hadn't sold rights to the film. Only evidence showing the sale predated the filming would prove a hoax.

Packham said of his first impression of the Patterson film, "I knew it was a hoax. Everyone did." This is another overstatement. Experts who viewed the film were split on it. There were many people, including some qualified Ph.D.s like Grover Krantz and John Bindernagel, who have argued it was genuine. Packham's claim to have "easily" duplicated the film using an actor in an ape suit wasn't borne out by the photographs in the article. The Packham ape suit, covered in obvious synthetic-fiber hair, looked transparently fake. If the Patterson film does show a man in an ape suit, it is, at the least, a very good ape suit.

The second attempt to bury the sasquatch is a long and complex tale. In 1999, a then-anonymous man from Yakima, Washington, speaking through his lawyer, announced that he had worn the costume to make the film. Sasquatch researcher René Dahinden scoffed, "This guy is probably number sixty-four who claims he was in the fur suit." The story, however, drew the interest of Seattle writer Greg Long, who started an investigation that would come to a head four years later.

Meanwhile, the end of 2002 brought another spate of "Bigfoot is Dead" headlines. These were sparked when a California man named Ray Wallace died in November at the age of eighty-four. His family claimed he had created the entire sasquatch legend by hoaxing the first well-known sasquatch prints, the ones which gave rise to the term "Bigfoot." These were found in northern California in 1958 by a road-building outfit belonging to Wallace's construction company and led by one Jerry Crew.

Sasquatch hunters had long known Wallace as a gadfly who enjoyed carving huge wooden feet and making dubious sasquatch movies. Not surprisingly, cryptozoologists were nowhere near as excited as reporters seemed to be about Wallace's claims. The carved feet displayed by Wallace's family were not a match for the footprint casts from 1958. Wallace also claimed to have been involved in making the famous 1967 Patterson-Gimlin film, but he offered no evidence of that.

In 2004, Greg Long published a book that attempted to bury the Patterson-Gimlin film once and for all. Not only was the man who had claimed to wear the suit—Bob Heironimus, a longtime acquaintance of Patterson's—revealed, but so was the man who claimed to have made the suit, professional costume maker Philip Morris.

Long's *The Making of Bigfoot* unquestionably demonstrated there were contradictions and other problems with the stories Patterson and Gimlin told. The book also showed that no thorough, independent investigation of the film story had been done when the evidence was fresh. Basic facts such as who developed the film and when had remained unclear.

Sasquatch enthusiasts, though, pointed out that Long's story had problems of its own. One was that he could not produce the suit, nor had he been able to discover what happened to it. Another was that Heironimus described a smelly, heavy home-made suit of horsehide, while Morris talked about a suit made entirely of a synthetic fiber called Dynel. The two stories just didn't match. Long wrote that Patterson had made some modifications to the Morris suit, but that didn't seem nearly enough to reconcile the descriptions.

Sasquatch sightings have not dropped off since the Wallace story or the Long book. Today, enthusiasts continue to scour the woods for footprints, hair, droppings, and the hoped-for sight of an unknown North American ape. For something that has been killed off repeatedly, sasquatch is pretty lively—be it animal or legend.

THE ENDURING "SEA SERPENT"

Since the beginning of recorded history, seamen and other witnesses have been reporting huge, elongated marine animals that don't fit into zoological textbooks. There is no doubt the majority of these people have been mistaken. A few have been hoaxers, although these often give themselves away by adding too many colorful details. Without physical evidence or good photographs, though, skeptics have consigned the sea serpent to the status of a legend, albeit a timeless and endlessly fascinating one.

Beyond the shadow of the ship
I watch'd the water-snakes:
They moved in tracks of shining white
And when they rear'd, the elfish light
Fell off in hoary flakes.

— COLERIDGE, The Rime of the Ancient Mariner

The most interesting thing about this subject is the way sea serpent reports have persisted from long before the days of Coleridge through the modern age of science. The longevity of the sea serpent phenomenon requires that the subject be reexamined to see if there could be a real animal unknown to zoology still lurking in the depths of mistake, legend, and myth.

The first question is whether it's still plausible that the oceans hold large animals we don't know about. The answer is a very solid "yes." In addition to the examples I cited in reference to the sighting by Nicoll and Meade-Waldo, there is the enormous fish seen at close range from the submersible *Deepstar 4000* in the San Diego Trough in 1966 and still unidentified. The two witnesses on the submersible described their visitor as thirty feet or more in length, with a rounded tail like a grouper's and "plate-sized" eyes. Another important example is Mesoplodon "Species A." Here we have a sizable beaked whale seen many times and even photographed, but no one has captured a specimen, and no confirmed dead examples have drifted ashore. (One point often made against the existence of sea serpents is that no carcasses have been found, so Species A is an important example to keep in mind.)

One crucial point to make is that any marine "monsters" which may exist are definitely not serpents. There are no reliable reports of such things moving with a snake's distinctive side-to-side motion. Modern sea snakes never exceed ten feet in length, although *Pterosphenus schucherti*, which became extinct in the late Eocene, may have been twice that size, and the similarly dated *Palaeophis colosseaus* even larger. The common usage has been with us so long, though, that we are pretty much stuck with the term "sea serpent" for now.

There is no good photographic evidence of sea serpents, with the possible exception of a case from 1937 (more on that later). Two videotapes, one made in Chesapeake Bay in 1982 and the other in June 1999 in Norway's Alesund fjord, do appear to show large serpentine animals. Unfortunately, even though scientists at the Smithsonian agreed the video of "Chessie" showed a swimming animal of some type, both videos were taken from too great a distance for positive identification. A creature filmed off Cornwall in 1999 by John Holmes, a former natural history museum curator, raised some excitement in the cryptozoological community. Today, though, it is generally agreed this was a sea bird, probably a cormorant. The case is a good reminder that judging sizes and distances of animals in a featureless ocean is a most uncertain art, even for someone with a related professional background.

Nor are there any known sea serpent bones, scales, or other hard evidence. Carcasses washed ashore and alleged to be sea serpents have either been identified as known animals or lost before they could be examined by experts. For a long time, a larval eel, or leptocephalus, six feet long, collected in 1930, was cited as proof of a gigantic adult. Most ichthyologists, though, now consider this specimen interesting but not extraordinary. It's not an eel at all, but one of a group of eel-shaped fish called the noticanthiforms (spiny eels) in which the larvae are as large or larger than the adults.

Accordingly, the credibility of the sea serpent largely rests on eyewitness accounts. Such accounts cannot be said to "prove" the existence of a new animal. What they can do, if convincing enough, is make this subject a legitimate one for further inquiry.

With this background in mind, let's examine some of the more prominent accounts. To avoid the problem of judging relatively ancient history, we will start with 1817. In that year, a "chocolate colored" snakelike creature said to be forty to a hundred feet long was seen in and around the harbor at Gloucester, Massachusetts, by over 100 witnesses. As June O'Neill showed in her well-researched and enjoyable book *The Great New England Sea Serpent*, sightings of this or a similar creature actually began in 1638 and have continued into modern times, but the summer of 1817 was the crescendo of this puzzling activity.

Minor details varied, but all the witnesses reported an animal which carried its head out of the water, had several "coils" or "humps," and swam with a marked vertical undulation at impressive speed. A local naturalists' group, the Linnaean Society, went about gathering reports and deposing witnesses in a most exemplary fashion. Unfortunately, the Society soon destroyed all credibility for itself and the serpent by identifying a deformed blacksnake with humps as the monster's young. Thus the sea serpent lost the best chance it ever had to attain scientific respectability.

Despite this, sightings in the Gulf of Maine and adjacent waters continued almost annually throughout the 19th century and occasionally still occur. In April 2000, a motorist on a coastal road reported seeing an animal with a thirty-foot body near Cape Bonavista, Newfoundland. The startled witness, Bob Crewe, said, "It looked something like a rock in the water, but I knew there was no rock there. I blew the horn and it stuck its head out of the water. It had a long neck about four or five feet."

The ships of Britain's nineteenth-century Royal Navy seemed to have a knack (welcome or not) for encountering unknown marine beasts. Captain James Stockdale of H.M.S. *Rob Roy* filed an impressive report with the Admiralty in 1815. His ship was off St. Helena when what looked like a gigantic snake with a dorsal fin passed so close Stockdale reported a strong "fishy" odor. He could also measure the animal against the ship. Stockdale claimed the sea serpent was over 120 feet long and held its head about six feet out of the water. He noted the beast was motionless until the ship came close aboard, when it seemed to be startled and swam off at about five knots.

The most famous sea serpent report of all time was made by the crew of H.M.S. *Daedalus* in 1848. In the south Atlantic, near the Cape of Good Hope, Captain Peter M'Quhae and his officers encountered a huge long-necked beast which, in an interesting contrast to the *Rob Roy* creature, completely ignored the nearby ship. The witnesses reported seeing the mouth and the eye of the creature. According to M'Quhae, the animal passed close enough that, "if it had been a man of my acquaintance, I would easily have recognized his features with the naked eye."

Writer Richard Ellis suggested in his 1994 book *Monsters of the Sea* that this animal was a giant squid. For this to be correct, Captain M'Quhae must have been considerably in error. There is some chance this may have been the case. In 1995, Ellis received a letter from a Mr. Maldwin Drummond. Maldwin is the grandson of Lieutenant Edgar Drummond, the watch officer on the *Daedalus*. Enclosed with the letter was a drawing from Edgar Drummond's journal. This drawing showed less detail on the animal and placed it much farther from the ship than appeared to be the case from the famous picture drawn under M'Quhae's supervision, which was published in the *Illustrated London News* and has been reprinted in every book on the subject ever since. Maldwin Drummond called the squid theory "very convincing."

Edgar Drummond's drawing also includes two objects well behind the "head" labeled "fins." Ellis thinks these "fins" might be tips of tentacles flailing above the water. There is one problem. Drummond's sketch shows only a small eye, not the huge eye of a squid. Overall, though, the drawing resembles a squid jetting along with its front end held just out of the water—if squid ever do this.

Diver with one of many renditions of a sea serpent.

If this was a squid, it must have been among the largest on record. M'Quhae originally thought the creature might be 120 feet long, while Drummond talked him down to a still-impressive sixty feet. The largest universally accepted specimen of Architeuthis, the squid stranded at Thimble Tickle, Newfoundland in 1878, was fifty-five feet long. British biologist Michael Bright, in his 1989 book *There are Giants in the Sea*, presented reports of squid from 72 feet to over 100 feet long, but documentation is lacking. As a final note on this detour into the shadowy world of the giant squid, writer Peter Benchley once heard from a prominent teuthologist who asked not to be named that "a lifetime of study had convinced him that the existence of a 150-foot giant squid was not only possible but probable." A creature of those dimensions would be a "sea monster" by anyone's definition.

A year after the *Daedalus* encounter, Captain George Hope of the H.M.S. *Fly* recorded sighting a strange animal seen in its entirety through the clear waters of the Gulf of California. The witnesses compared the beast to an alligator, except it had a very long neck and fins instead of feet. The officers of the British royal yacht *Osborne* described a gigantic marine animal in the Mediterranean in 1877. This creature was swimming away from the ship with its head out of water, paddling with front flippers the witnesses estimated were fifteen feet long.

In 1892, Dutch zoology professor A. C. Oudemans presented 187 reports in his book *The Great Sea Serpent*. Unfortunately, this well-qualified authority's belief that all reports concerned an enormous long-necked pinniped led him to twist some reports to fit his theory and to propose as the sea serpent a most implausible-looking animal with a gigantic tail.

In 1893, the crew of the steamer *Umfuli* reported a plesiosaur-shaped creature in the south Atlantic with a total length of about eighty feet. That figure included a fifteen-foot neck which carried the animal's head (which the *Umfuli's* mate compared to the head of a giant conger eel) well above the water. The creature had the body circumference of a "full-sized whale." Captain R. J. Cringle wrote later, "I have been told that it was a string of porpoises, that it was an island of sea-weed, and I do not know what besides. But if an island of sea-weed can travel at the rate of fourteen knots, or if a string of porpoises can stand fifteen feet out of the water, then I give in, and confess myself deceived. Such, however, could not be." He also made the weary comment that, "I have been so ridiculed about the thing that I have many times wished that anybody else had seen that sea-monster rather than me."

From 1897 to 1904, a series of encounters took place between

French Navy ships and apparent sea serpents in the coastal waters of Indochina. Several warships reported encountering elongated black-and-gray animals which swam with vertical undulations. The creatures had crests or fins (described as "saw-toothed") on their backs. Their heads were shaped like those of seals. Estimates of the animals' length ranged from 60 to 100 feet. The gunboat *Avalanche* pursued and fired at these creatures (which were often seen in pairs) on two occasions, but the animals outdistanced the ship. Some of the reports filed by naval officers stated specifically that the creatures never spouted and were not whales. This episode remains one of the more puzzling events in sea serpent history.

The most intriguing sea serpent report is still the 1905 encounter by Nicoll and Meade-Waldo. Having already presented this episode in some depth (so to speak), there's no need to repeat the details here.

Sea serpent sightings in the new century continued. In 1917, the officers of H.M.S. *Hilary* reported an animal with a long neck and a triangular dorsal fin. As seen from thirty yards away, it displayed a head shaped like a cow's, with a white marking running down the center of its face. The fin seemed "flabby" and cocked over at the top, a seemingly mammalian feature. Unlike the *Daedalus'* serpent, this creature lifted its head and turned its neck to follow the passage of the ship. According to the account by the *Hilary's* Captain F.W. Dean, however, studying the ship was not a good idea on the sea serpent's part. This being the height of World War I, Dean ordered the animal used for target practice. One six-pounder gun scored a hit, causing the thing to submerge in a welter of foam and spray.

Dr. Bernard Heuvelmans analyzed over 350 sightings he considered valid in his 1968 book *In the Wake of the Sea Serpents*. Heuvelmans concluded there were at least seven new species involved (five mammals, a reptile, and an eel). Understandably, Heuvelmans' fellow zoologists found the idea of a whole zoo of huge unknown animals every bit as hard to swallow as Oudemans' one-size-fits-all pinniped.

In 2003, marine biologist Bruce Champagne completed a new study of sea serpent reports. He sifted through no less than 1,247 reports of large unknown marine creatures, accepting 358 as usable. He found that even the apparently reliable data covered a very broad spectrum of descriptions: indeed, he needed ten categories of animals, not including the strays which fell into the classes "Unidentified" and "Other." Some reports even indicated subcategories. Champagne's top four groups, measured by the number of reports, were designated the Type 1 Animal (Long Necked) (only seven percent of the 358 reports, surprising given the traditional picture of the sea serpent as a long-necked creature), Type 2 Animal (Eel-like, nine and a half percent), Type 3 Animal (Multiple Humped, the most common with twenty-nine percent of the reports) and Type 10 Animal (unknown cephalopods).

Champagne is not arguing that all the resulting categories (fourteen of them, counting subcategories) represent genuine animals. What his work has done is to highlight the difficulty of getting any definitive identification out of the data, as even the most exhaustive sorting of reports to date essentially puts unknown sea animals all over the zoological map.

Champagne does think there is something to the sea serpent business. He has theorized that the Multiple-Humped category in particular is likely to represent at least one real animal. He suspects this animal is a reptile, albeit one with some adaptations to allow its range to extend into relatively cold waters. The animal appears to like coastal waters and estuaries. Such a creature could be responsible for the New England sightings, Caddy, and perhaps some "lake monster" reports if it traveled upriver and established itself in fresh water.

Champagne further suggests that the discrepancy between the classic picture of a sea serpent as long-necked and the relatively low number of truly long-necked animals which emerge when the reports are studied further could be due to animals "spyhopping." Orcas often spyhop, thrusting the head and forebody well out of water to survey what's happening above the surface. A relatively slender animal spyhopping would look, at first glance, like a long-necked creature.

Sea serpent reports have declined in the modern age of steamships and skepticism, but they have certainly not ceased. For example, there were several sightings from the California coast north of San Francisco in October and November of 1983. The most intriguing description came from a road-building crew. While working atop a seaside cliff, the workers saw a huge, elongated creature in the surf directly below them. An engineer using binoculars estimated it was black, eel-like (though undulating vertically), and as much as 100 feet long. In 1997, two fishermen in Fortune Bay in the Gulf of Maine reported they had come within sixty feet of a sea serpent. The creature raised a horselike head, with visible ears or horns, on a six-foot neck. The overall length appeared to be thirty to forty feet. The Welsh port of Milford Haven hosted an intriguing sighting in March 2003, when several patrons at a waterfront pub described watching a dark, elongated animal with a snakelike head and a dorsal fin swim past at a good clip before diving out of sight. A traditional British hot spot for sightings is off

Basilosaurus, an acheocete some theorize is behind stories of sea and lake monsters.

Falmouth, where a dark-skinned beast with a small head and one or two humps was reported twice in 1999 and once in 2000.

In June 2003, a lobsterman named Wallace Cartwright was fishing off Nova Scotia when he and his crewman saw an animal about seventy yards away. They followed it for forty-five minutes as it swam toward the open ocean at three or four knots, diving and submerging five or six times. According to Cartwright, a thirty-year sailor familiar with all types of marine life, the thing had a head like a sea turtle, which it sometimes raised out of the water, and a smooth, brownish body over twenty-five feet long. Andrew Hebda, a zoologist at the Museum of Natural History in Halifax, suggested Cartwright might have seen an oarfish. Cartwright looked up oarfish and rejected any similarity between that distinctive red-crested, silver-sided rarity and his animal. "I don't know what it was, but I know what it wasn't," he declared.

Reports off the northwest coast of North America are, if anything, increasing. The first non-Native mention of an unknown animal in this region was recorded by an American fur trader, Robert Gray, in 1791. Over 200 years later, Canadian scientists Paul LeBlond and Edward Bousfield formally described the animal involved (nicknamed "Cadborosaurus" or "Caddy") as a reptile, *Cadborosaurus willsi*. Their type-specimen was a "juvenile" carcass found in a sperm whale's stomach in 1937. This was removed intact at Naden Harbor, British Columbia, but eventually lost.

Photographs which survive show a very slender animal, about twelve feet long but only six inches in diameter. Front flippers are clearly visible, and the mangled-looking remains of what appear to be hind flippers or tail flukes hang down at the rear. The neck is elongated, although not to the extent often described by Caddy witnesses. The head is vaguely camel-like. The whalers and newspapermen who saw this thing thought it very strange, and their accounts say it was not decomposed. Samples were reportedly sent to a museum and were identified as part of a fetal baleen whale. Bousfield and LeBlond object that 1) the photographed object does not in the slightest resemble a whale fetus, and 2) there is no logical way such a fetus could end up in a sperm whale, which has never been known to prey on baleen whales in any manner.

According to Bousfield and LeBlond, adult Caddys may be fifty feet long, making this species the longest of all living reptiles. Caddy displays considerable vertical flexure, often resulting in what witnesses describe as loops of its body. It has large eyes and four flippers. The animal can swim at speeds reportedly as high as forty knots. In a unique report, two pilots in 1993 claimed to have seen a pair of Caddys from above in Saanich Inlet. It's been speculated the aircraft interrupted a mating encounter.

The authors' identification of the animal as a reptile was almost as controversial as the claim it existed at all. Bousfield and LeBlond also relied on two reports, both from the testimony of single witnesses and one quite vague, of close-up encounters with what they think were baby Caddys. These were very small (about eighteen inches long), and both were described as reptilian.

Even if one accepts the 1937 photograph shows an unknown species, it does seem the Canadian scientists overreached the available data in assigning the creature to the reptiles. It's difficult to interpret eyewitness reports, and the "baby" stories are unverified.

There are other problems with the reptile theory, the most significant being that there are no known cold-water reptiles. The massive, slow-swimming leatherback turtle (*Dermochelys coriaceacan*) can visit these latitudes, but it has a totally different body type. If Caddy was a true reptile, it would be relatively sluggish, at least part of the time, and not the very fast swimmer often reported. Additionally, it would have to spend much of its time basking on the surface. Such a creature in this region would be easy prey for killer whales.

When I offered this opinion to Dr. LeBlond, he wrote back, "We had long arguments before opting for the reptilian classification and finally decided to be more definite than not in order to focus the discussion. If this was a matter of betting, I wouldn't risk my money; however, it is a matter of discovery and scientific discussion, an area where one can be a little bolder."

Two herpetologists raised the above objections in a 1996 article in the journal *Cryptozoology*. Aaron Bauer and Anthony Russell added their judgments that Caddy, as described by witnesses, showed almost no reptilian features and that the degree of vertical flexure attributed to *C. willsi* was impossible for a reptile. The authors suggested the Naden Harbor carcass, witness descriptions notwithstanding, was most likely a decayed basking shark. The authors did admit there are no records (or, at least, no other records) of sperm whales swallowing basking sharks.

Sea serpents, whether they are animals or only legends, will be with us for a long time. Unlike the mysteries involving lake monsters or many land-dwelling cryptids, which could be solved definitively one way or the other given sufficient resources, sea serpents have the enormous volume of the Earth's oceans to hide in.

Sea serpent sightings are scattered all over the world. Several of the classic cases come from the south Atlantic, while the New England saga offers the greatest number of sightings in one area and the Cadborosaurus case involves the largest number of comparatively recent sightings. (LeBlond counted fifty-three Caddy sightings between 1812 and 1984, twenty-three of which he classed as

One of the first Loch Ness sightings, reported by veterinary student Alan Grant.

unexplained.) An air and sea search off the New England and British Columbian coasts would seemingly offer the best chance of success, but even so, a lot of luck would be needed to prove the animals' existence. Given that situation, proving the nonexistence of sea serpents is a devilishly difficult proposition. If sea serpents are real animals which do not breathe air—for instance, if they are a species of deep-water eels which surface only on rare occasions—they could hide for another century.

The case for the sea serpent is not proven, and won't be until there's a fresh carcass. However, there is a good chance, based on the most authoritative reports, that the seas do hold at least one large unknown animal. The one that seems most likely is a giant eel or eel-like fish. There might even be two animals, with the less definite (but intriguing) possibility being a long-necked animal, perhaps a mammal of some sort. What sort? We probably won't know until we catch one.

LOCH NESS: WHAT'S GOING ON?

Easily the most famous cryptid of all time is the reported inhabitant of Loch Ness in Scotland. Thirty years ago, "Nessie" often made the newspapers and had numerous backers in the scientific community. As the new millennium dawned, though, news from Loch Ness was getting short shrift at best.

Has anything really changed? Have Nessie reports dropped off, or is it only that the public (and most interested scientists) have grown weary of seeing more of the same old evidence without any new revelations?

The Nessie business really began in the 1930s. There are only a handful of earlier reports, none of them well-documented. The oldest report of all, a colorful account of Saint Columba confronting a monster in the year 565 or thereabouts, is actually placed in the River Ness, which connects the loch to the sea. The existence of this river suggests the possibility of an ocean-dwelling creature either migrating in and out or trapped in the loch. Unfortunately, the river is very shallow today, and nothing larger than a seal could make such a journey undetected. (In fact, a seal did wander into the loch in 1984. It was pointlessly shot, making one wonder whether the animals or the humans are the real "monsters" here.) The loch, for the record, is about twenty-four miles long, up to 900 feet deep, and has a substantial, though not bountiful, population of fish including eels, salmon, and char.

The major outbreak of sightings which occurred in 1933 may be linked to a road-improvement project along the north side of the loch, which provided motorists with a much better view and could conceivably have disturbed a creature living in the water. Then, as now, there was a tendency to refer to "the monster" as an individual. To explain the record of sightings over the next sixty years, however (given the aforementioned lack of access between the loch and the ocean), a breeding colony is required. Incidentally, the loch surface is some 50 feet above sea level, ruling out any connection to the sea via an underground cavern. (Claims of finding cave or tunnel entrances on sonar have been made, but never verified.)

From the beginning, some people have suggested the whole thing was a hoax. One of the first well-publicized sightings, in May 1933, was made by the couple who ran the lakeshore Drumnadrochit Hotel. The sighting was reported to the *Iverness Courier* by Alex Campbell, a water bailiff (an official who enforces fishing laws). Campbell later claimed to have made several sightings himself, and some skeptics believe he virtually invented the monster. Another man, a local author who wrote under the name Lester Smith, once told investigator Henry Bauer that he invented Nessie to draw business for local hotels.

What were people reporting? Sometimes it was merely a hump, or two or more humps in line. Some people (such as Campbell, in his own first sighting in September 1933), reported a large hump plus a slender neck, estimated to stretch five feet long or longer, and a small head. The whole visible length, Campbell thought, was about thirty feet. According to the first on-site survey of witnesses, conducted by Rupert Gould in 1933, this plesiosaur-type configuration matched most sighting reports. The chief discrepancy between witnesses concerned the number of humps.

Eyewitness reports since then have added virtually nothing to this basic picture. Witnesses differ (as might be expected) on the presence and shape of small features such as eyes, "horns," etc. The color has been described as everything from black to reddish-brown to olive drab, only rarely with mottling or other markings.

KEY DATES IN NESS HISTORY SINCE 1933:

1934: Dr. Robert Wilson reports taking the "Surgeon's Photograph," actually two pictures, the first showing a classic plesiosaur-type head and neck shape. Skeptical suggestions concerning the shape's identity include a hoax, the tail of a diving otter (suggested by the late Dr. Maurice Burton, a leading zoologist), and a waterbird (the opinion published by Dr. Roy Mackal, a believer in Nessie).

1957: Constance Whyte publishes one of the most famous Loch

Ness books, More Than a Legend, including sighting reports and the seven photographs then available.

1960: American engineer Tim Dinsdale shoots a 16mm film of what looks like a humped back moving through the loch. The hump is moving across the lake at first, then turns and moves parallel to the shore, at one point appearing to submerge. Subsequent enlargement shows what appears to be a second, smaller hump appearing momentarily behind the first. Skeptics suggest the object is a boat, its features obscured by distance.

1968: Researchers from the University of Birmingham obtain sonar records of what appears to be a large, solid object, rising and diving at speeds too great for schools of fish.

1972 and 1975: Expeditions from the Academy of Applied Science (AAS), a Boston nonprofit foundation established by attorney Robert Rines, obtain more sonar records, along with the first underwater photographs. When the photographs are enhanced, they appear to show a diamond-shaped flipper, the body and neck of a plesiosaur-type animal, and a close-up of a head with a blunt muzzle and some type of horns or knobs on top. Some skeptics have subsequently accused Rines of retouching the pictures, which he denies.

1977: Gwen Smith films a polelike object which rises from the water to an estimated height of six feet, then submerges. The camera is too distant to show detail. Writer Steuart Campbell suggests Smith was a victim of a hoax, although the evidence is circumstantial.

1987: The Loch Ness and Morar Project (founded to investigate Nessie as well as Morag, the similar creature reported in Loch Morar) mounts Operation Deepscan, in which a line of boats attempt to deploy a sonar "curtain" the width of the loch and sweep it from end to end. The first day's work produces three contacts of interest, one apparently the size of "a large shark." A second day produces nothing. The expedition does photograph a submerged tree stump which Project leader Adrian Shine believes to be the "head" photographed by the AAS.

1993: A ninety-year-old man named Christian Spurling claims the Surgeon's photograph was a fake, a model he built on a toy submarine. Spurling's stepfather, a filmmaker and big game hunter named "Duke" Weatherell, was allegedly behind the hoax. Skeptics proclaim the monster dead: some believers point to numerous inconsistencies in Spurling's story (he died before his story was published, preventing follow-up questioning) and argue the confession itself is the hoax.

1997: Salesman Richard White takes a series of still photographs near the village of Foyers and wins a prize of 500 pounds from the bookmaking firm William Hill for the best Nessie photo of the year. (The bookie continues to offer odds of 250-1 against anyone producing physical evidence.) The photographs show a round object apparently making a V-shaped wake: they are interesting, but don't offer enough information to identify the object.

1998: A vacationer named Geoff Mitcheson videotapes the head of an animal. Gary Campbell of the Loch Ness Monster Fan Club declares the tape the best evidence ever for the monster, but a zoologist, Dr. David Waugh, says, "We have a very high level of confidence that this was a seal."

1999: The first sighting of Nessie via Internet is announced. Two enthusiasts in Texas, watching a Webcam positioned at Loch Ness, believe they spotted the elusive "monster" of the loch on June 5, 1999. The video still confirmed there was a large dark object in the water, but no more.

2001: New photographs taken by James Grey are very sharp and clear, but the long, slightly curved object shown has no detail. While it moves from one picture to the next, it never flexes. Some researchers suspect it's an inanimate object. Nessie-hunter Andreas Trottman, while investigating this and other recent events, also finds a new hoax: two dead conger eels, seven feet long, apparently planted on the shore of the loch.

2003: Yet another hoax surfaces when a fossilized plesiosaur vertebrae is found on the Loch shore. However, the limestone in which it is partly embedded shows it came from a formation at least thirty miles from Loch Ness. A few months later, the BBC sponsors another sonar survey. Nothing convincing turns up.

Without definitive evidence for the existence of Nessies, it is not very useful to speculate on what they might be. Despite this fact, no one involved in the subject can resist such speculation. Some of the suggestions made so far include giant otters, huge thick-bodied eels, embolomers (greatly enlarged descendants of prehistoric amphibians), plesiosaurs, archaeocetes (primitive whales), other types of mammals, including hypothetical long-necked sirenians or pinnipeds, and invertebrates of various kinds.

There are countless factors thrown into the arguments over what might and might not be reasonable candidates. The water in the loch below the sun-warmed surface is only a few degrees above freezing, which argues against a reptile. The many reports of Nessie swimming at high speed also rule out the long-necked types of plesiosaurs, which were clearly not built to zoom through the water. There is also considerable difference of opinion on how flexible and maneuverable a plesiosaur's neck was. It's now considered unlikely a plesiosaur could actually present the kind of "classic plesiosaur" silhouette shown in the Surgeon's Photograph.

The rarity of the creatures' appearances at the surface makes it harder to accept mammals or any other air-breathers. The reportedly variable geometry of the creature's "humps" is difficult to assign to any type of vertebrate. On the other hand, no known invertebrate (except the cephalopods, which are clearly not involved) comes close to the required size. Loch Ness is not a rich lake from a biological viewpoint, and a colony of large predators should be doing serious damage to the fish population. There's no evidence that's taking place.

The Loch Ness monster, real or not, is an established celebrity. There are countless Web sites, at least two with live cameras trained on the loch, and several organizations, such as the "Official Loch Ness Monster Fan Club." (One wonders how any such enterprise becomes "official"—are they claiming Nessie's endorsement?) At Loch Ness itself are two exhibitions and plenty of opportunities to buy stuffed monsters, vials of Loch water, photographs, and even "monster droppings" (lumps of peat). By one estimate, the monster is worth twenty-five million pounds a year to the local economy.

Tim Dinsdale, after decades living at and on the loch in pursuit of his quarry, died in 1987, but other determined researchers are still trying to obtain proof of large unknowns in Loch Ness. Adrian Shine still directs what's now called the Loch Ness Project. Robert Rines has a home on the loch and is still mounting periodic expeditions with his organization, MonsterHunters.org.

As mentioned earlier, press attention has abated considerably since the 1970s, but it certainly hasn't disappeared. Neither have sightings, although one tally, at least, shows a drop-off. Twelve encounters were reported to Campbell's group in 2000, but only four in 2001 and three in 2002. The following year showed an uptick: the first day of June 2003 brought a record three sightings reported in one day.

At this writing, the skeptics have the advantage in the Nessie debate. The continuing lack of physical remains or clear, close-range images has tilted the balance of scientific opinion heavily in their favor. This does not mean the argument is over. If there are Nessies, then sooner or later some lucky researcher or tourist will get the film to prove it, and science will be much the richer. If there are no Nessies, the study of the lake has still produced a great deal of data on its ecosystem, fish life, and so on, so the effort will not be wasted.

I've become very skeptical myself, although I can't quite dismiss the subject completely. The object in the Dinsdale film does not look to me like a boat, and the various enlargements and enhancements have failed to bring out details supporting the boat hypothesis. That, along with some of the sonar evidence, is just enough to prevent me from closing the book on this case.

Despite the frustrating lack of evidence and the numerous problems from a biological viewpoint, it seems there will always remain an aura of mystery around Loch Ness. This may not be very satisfying for science, but it is, I think, good for the human spirit. A world without mysteries would be a grim place indeed. May Nessie (whatever it is) live on!

THE STATE OF THE SASQUATCH

The question of a large unclassified primate in North America has gone through stages similar to those encountered by the yeti and the Loch Ness puzzle. Thirty years ago, the subject was seriously discussed at scientific gatherings and prominently reported in the news media. At this writing, the level of interest appears to have declined. Many once-curious anthropologists and primatologists have dismissed the subject. It's worth taking a fresh look to see what the sasquatch situation is today and how it came to be.

The origins of the sasquatch saga are shrouded in controversy. The term itself comes from a Salish Indian word. Various Native American tribes, mainly in the Pacific Northwest but also further east and south, told stories of large, sometimes hair-covered beings living in the forests. While many tribes, especially the hunting cultures, had a thorough knowledge of native fauna, it's often difficult for the non-Indian to be certain whether a particular tradition refers to animals, supernatural beings, or something between the two.

Modern reports of an apelike animal in the northwestern U.S. and western Canada began to circulate from the 1920s on, but the first big blaze of publicity came with the Crew case of October 1958. Jerry Crew, part of a road-building gang in northern California, was photographed holding a plaster cast of an enormous footprint found near his work site. Newspapers around the world published the picture.

The most controversial Bigfoot evidence came from the same region in October 1967, when Roger Patterson and Bob Gimlin filmed a hair-covered biped walking through the forest along Bluff Creek. The 16mm film showed a heavyset animal moving with a fluid, long stride diagonally away from the two men, who were as close as a hundred feet at one point.

The film was impressive enough to be reviewed by "mainstream" experts who previously ignored the subject. One such was Dr. John Napier, one of the world's leading primatologists. Napier's views were mixed. If the film was a hoax, it was an impressive one. If not, it showed an animal which, in Napier's opinion, presented a highly improbable mix of human and ape characteristics (human legs and gait, but a massive upper body) and male and female traits (a sagittal crest like a male gorilla, but prominent breasts like a human woman). On balance, he finally rejected it as bogus.

An incredible amount of effort has been expended arguing the details of this film over and over. Suffice it to say that if there has ever been an authentic sasquatch film taken, this is it. If it's a hoax (either perpetrated by Patterson and Gimlin or on them), the hoaxer knew how to make a convincing ape suit. Hollywood makeup artist

Sasquatch, North America's problematic primate.

John Chambers, who created the costumes for the *Planet of the Apes* movies, was a natural suspect, but Chambers flatly denied any involvement. He told an interviewer, "I'm good, but not that good."

Grover Krantz wrote in his 1999 book *Bigfoot Sasquatch Evidence* that he had seen eight purported sasquatch films made since 1968. All of them, he was certain, were fakes, but he was convinced Napier was incorrect in rejecting the Patterson film. The seemingly endless point-counterpoint debate over this film will undoubtedly go on until there is another film of equal or better quality (or, of course, a real sasquatch) to compare it to. The Patterson film is probably the most analyzed amateur film in all of history, with the possible exception of the Zapruder film of U.S. President John Kennedy's assassination. The tiny sasquatch image has been enlarged, enhanced, studied frame by frame, turned into computerized animation of what the skeleton may have looked like and how it moved, compared to footprints from the film site and other locations, etc. Almost every time a new study is done, the people involved believe they have found some new details that either prove or disprove the film's authenticity. Without plunging into the details, which demand a book in themselves, two things should be mentioned. One is that no one has yet pointed to an irrefutable artificial detail like a zipper. The other is that photographic interpretation, for all the tools of the 21st century, remains a bit of a subjective art. The certainty of American intelligence analysts about Iraq's illegal weapons, a primary rationale for going to war with Iraq in 2003, offers the caution that the finest analysts, using the best technology, can still fall prey to the human failing of seeing what one wants or expects to see.

In 1969, evidence was discovered which some experts found more convincing than the film. This was a trail of sasquatch tracks near Bossburg, Washington. The tracks were huge, eighteen inches in length. The right foot showed an anatomically accurate clubfoot deformity. Napier wrote, "It is possible to conceive of a hoaxer so subtle, so knowledgeable ... who would fake a footprint of this nature ... but it is so unlikely I am prepared to discount it."

In the last thirty years, a number of alleged sasquatch films and photographs have been proffered as evidence, along with innumerable sightings and footprints. None, not even the footprints found in 1982 and 1984 which show dermal ridges, have had the impact of the Patterson film and the Bossburg prints. (Those prints have been alleged to be clever fakes, although experts disagree over the plausibility of faking footprints with dermal ridges.)

A splash was made in 1995 by reports that purported sasquatch hair would be tested at Ohio State University using a new DNA analysis technique. However, Dr. Frank Poirier of the university's anthropology department said no such testing was done and the news report "should not have been written."

While both key pieces of evidence are worth studying, neither is unequivocal. The main problem with the Patterson film is the division of expert opinion concerning whether the creature involved looks plausibly real. The chief investigator of the Bossburg tracks was longtime sasquatch hunter Ivan Marx, who has since claimed success in filming the very sasquatch involved. The authenticity of Mr. Marx's film is, to put it charitably, doubted by almost all cryptozoologists. There is no proof Marx faked the Bossburg tracks, but his involvement in making and selling the questionable film raises suspicion. Grover Krantz believed Marx's early evidence was genuine, but, after years of frustrated efforts to obtain more convincing proof, Marx turned to faking it.

Soon after Chris Packham's assault on the Patterson film came a claim by two sasquatch researchers, Cliff Crook and Chris Murphy. They announced that digitized enlargements of key frames showed an artificial object, possibly a fastener or zipper pull, about two inches long, near the animal's waist. This was widely reported in the media, but apparently not one reporter asked to see the actual enlargement before writing the story. Neither did said reporters seem to notice the contradiction in the idea of a hoaxer who would craft a very impressive ape suit and then leave a clearly artificial object hanging out of it. In any event, the enlargements turned out to show a fuzzy blob that could have been just matted fur, rather than anything definitely man-made.

The scientific problems with sasquatch begin with the question of origins. There is no fossil record of man-sized or larger primates in the Americas. Sasquatch partisans will no doubt point out the relevant fact that the fossil record of the African gorilla consists of exactly one canine tooth from Kenya, and there isn't a single widely accepted chimpanzee fossil from any location. So the lack of fossils is important, but it does not, by itself, damn the sasquatch to nonexistence.

As Kenneth Wylie has pointed out, there are other problems in accepting sasquatch as a primate. Among the known higher primates, none are solitary, although an overwhelming majority of sasquatch reports describe a lone individual. Many sasquatch reports indicate an animal that is often active at night. There are no nocturnal apes, and only a few species of nocturnal monkeys. This does not mean sasquatch can't exist, but, once again, it raises the standard for considering evidence.

While sasquatch has been reported all over the U.S., it is impossible to accept that a large, undescribed species has such a wide range. The most likely habitat remains the Pacific Northwest, which includes large tracts of still-virgin forest. An intelligent, cautious primate, with a total population perhaps in the hundreds, could avoid humanity for a long time. But could sasquatch forever avoid being hit by a truck, or shot by an elk hunter? The Florida panther, a wary animal whose population numbers only in the dozens, still suffers deaths from automobile collisions every year.

Another problem is food. There is plenty of plant matter available in this region, but it's of relatively low energy value. Napier went so far as to say the area is a perfect habitat for a large ape species—"if it never had to eat." Dissenters, such as Krantz, have argued the problem has been overstated and that an omnivorous ape would be all right. Ecologist Robert Pyle, in his excellent book *Where Bigfoot Walks*, agreed with Krantz on this point.

Krantz once made a useful distinction when asked if he believed in the creature. "I don't believe in Bigfoot," he said. "I have certain knowledge that causes me to conclude." Krantz went so far as to name the species of sasquatch, based solely on footprints. He called sasquatch a living representative of *Gigantopithecus blacki*, a huge ape known solely from fossil teeth and jaws found in southern Asia. Krantz's fellow scientists generally believed he went too far in naming sasquatch without a body in hand. Naming a species using only a track or footprint cast as the holotype is a common enough practice with extinct species, particularly dinosaurs, but is very rarely done with living animals. Not having any record of Gigantopithecus foot bones or footprints made Krantz's action all the more speculative.

Whatever the level of scientific interest in sasquatch, there remains a great deal of popular interest. Several organizations are investigating the sasquatch question. These run the gamut from balanced scientific inquiries to what might be called "fan clubs" willing to print any information as fact. (Robert Pyle wrote of some of the more rabid researchers, "These people don't want to find Bigfoot. They want to be Bigfoot.") Ray Crowe's Western Bigfoot society is one of the most active groups. Another is the Bigfoot Field Researchers Organization (BFRO), founded by Matthew Moneymaker and Ron Schaffner. (For contact information, see the INTERNET section.)

Sasquatch has found a place in modern American culture, where it is used to sell trucks, tires, and pizzas. The town of Willow Creek, California (fifty miles south of the most famous sasquatch haunt, Bluff Creek) boasts a famous Bigfoot statue, a Bigfoot Museum, and a road called the Bigfoot Byway. Creston, British Columbia, in 2004 announced plans for its own statue of a sasquatch, a bronze creation almost ten feet tall and toting a case of Kokanee beer (which is, of course, brewed in B.C.). The Seattle Supersonics basketball team has a mascot named Squatch.

Officially, the animal remains unrecognized. There were newspaper reports in 1978 stating the animal had been placed on the official U.S. government Endangered Species List, but the origin of this claim is unclear. When queried, the Fish and Wildlife Service replied that no such recognition had been extended to this "legendary creature."

What is lacking in the sasquatch saga is physical proof: a bone, a hunk of skin, anything. Hair and dung collected at various sighting scenes have so far failed to provide positive evidence of sasquatch. This is partly due to the inherent nature of such evidence. Expert analysis may be able to say a hair or dung sample is "unidentified," but cannot prove it's sasquatch without a known sasquatch sample to compare it to. The lack of specimens was the number one reason cited by skeptical anthropologists in a 1978 poll. Only 12.8 percent of the anthropologists who responded to this query thought sasquatch reports concerned a real animal unknown to science.

Some cryptozoologists believe physical evidence does exist—the frozen corpse of the "Minnesota Iceman," a carnival exhibit which two zoologists examined in 1969 and believed to be the genuine corpse of an unknown man-sized primate. As in the case of the Patterson film, debate has raged ever since concerning almost every point of this supposed being's anatomy. However, there are two reasons for leaving this episode out of the current discussion. First, the corpse is not available to examine, its present whereabouts being a complete mystery. Second, even if this thing is genuine and not another clever hoax, the six-foot Iceman does not match the witness descriptions offered for sasquatch, which is considerably taller, broader-shouldered, and hairier.

Sightings keep piling up. Indeed, 2000 was called a "banner year" by cryptozoologists. This was largely due to two events. The first was a brief but clear sighting reported by a psychologist, Dr. Matthew Johnson, on July 1 at the Oregon Caves National Monument. Johnson had never considered sasquatch's existence possible, but was so impressed by his encounter he has been looking for the beast ever since. The second event was the finding of what enthusiasts believed was a body impression where a sasquatch had sat down and leaned to one side in a muddy spot in southern Washington state. An impression of this item, known as the "Skookum Cast," has been called at least plausibly authentic by a few scientists, including Dr. Daris Swindler, an anthropologist well established as an authority on primate fossils.

Sasquatch sighting reports were logged well before the 1958 incident publicized in the recent stories on Ray Wallace and will no doubt continue. Unfortunately, more sightings don't add anything, either in terms of facts or of "proof," unless they lead to the aforementioned hard evidence. (The exception would be a multiple-witness sighting by qualified biological scientists, an event which has not occurred so far.) Efforts by Krantz, biologist John Bindernagel, and others to sift sighting evidence and derive profiles of the animal involved are not convincing to skeptics because no researcher can be certain which sightings (if any) involve sasquatch and which involve hoaxes or misidentifications.

The most pervasive evidence is the footprints. Unfortunately, the thousand-plus footprint cases reported show every conceivable variation, including anywhere from two to five toes. Obviously, most are hoaxes. Krantz argued a minority of alleged sasquatch tracks were genuine, and believed he had identified traits which distinguish real tracks from fakes. However, even Krantz was apparently fooled by a hoaxer on at least one occasion. Krantz also argued that the sheer number of footprint reports, and the remote location of many, argued against hoaxes being responsible for all of them.

A majority of biological scientists nonetheless dismiss sasquatch, or, at best, pay no attention to it. Russell Ciochon has flatly called it a myth, and Alan Rabinowitz declared, "It is very rare, once you've been told about an animal and its habits, to then never find anything tangible."

There have been exceptions. Grover Krantz was a respected scholar of human evolution (although his reputation suffered from his identification with sasquatch). John Napier's preeminence in his field was unassailable. Dr. John Bindernagel has been after sasquatch ever since he found a set of what he believes were genuine tracks in a park on Vancouver Island. Primate researcher Jane Goodall told an interviewer, "I'm sure they exist." George Schaller of the WCS has not endorsed the animal's existence, but suggested that, given the sheer volume of the evidence, "I think a hard-eyed look is absolutely essential."

Skeptics say it is not closed-minded to demand the same evidence for sasquatch that is demanded for any other new species—a type specimen, alive or dead. This need not be the whole animal, just enough tissue to allow for DNA analysis to determine the sample really is from a new higher primate.

Dedicated searchers are still trying to get that type specimen. Conservationist Peter Byrne is still hunting sasquatch after four decades, and Canadian journalist John Green has pursued the animal even longer. Anthropologist Jeff Meldrum is another expert who, basing his judgment mainly on footprint casts, believes it's worth spending his spare time tramping through remote forests looking for sasquatch.

Is there anything real for these searchers to find? As so often happens in cryptozoology, the answer may still be, "we don't know."

It is, in my opinion, very unlikely there is a breeding population of huge apes living unknown to science on the North American continent. The search has gone on too long without producing physical evidence. I'm not denigrating the witnesses, many of whom are unquestionably sincere. I just feel it's more likely the sincere ones are mistaken than it is that sasquatch exists.

I genuinely wish to be found wrong in this judgment. I hope sasquatch really is out there somewhere, watching with puzzlement and perhaps even amusement as we humans blunder around looking for him.

CREATURES OF THE LAKES

Loch Ness is the most famous lake with a "monster," but it's hardly the only one. Large unknown animals, either serpentine or plesiosaur-like, have been reported from over a hundred lakes worldwide. Most of these are from the Northern Hemisphere, but there are a few in South America and Africa as well.

Making sense of this subject is very difficult. It is obviously not possible that nearly every sizable lake in the world—or even a significant fraction thereof—houses large unknown beasts. If we assume most cases can be written off as the products of mistakes, folklore, or both, there are still several large lakes which have substantial bodies of sightings and even some photographic evidence. Let us briefly review two of the most famous cases and see whether there might actually be something worth looking for.

Outside of Loch Ness and Lake Iliamna (the subject of an essay in the next section), the most interesting body of reports comes from Canada's Lake Okanagan. Lake Okanagan is the largest lake in the province of British Columbia. It somewhat resembles a larger Loch Ness, being seventy-nine miles long and two miles wide, with a maximum depth of 800 feet. The lake's creature, known as Ogopogo (not a name calculated to enhance scientific respectability), has allegedly been around since humans first arrived in the region. Native Americans who had to cross the lake supposedly threw in an expendable animal to keep the monster they called *Naitaka* busy so it wouldn't go after their canoes.

There are many modern sightings of something long and sinuous poking its head and back out of the water. Ogopogo reportedly has a horselike head on a visible neck, although the neck is not as slender as in Loch Ness reports. Many reports include "ears" or "horns" on top of the head. Dark green and black are the colors most often reported. The overall length of the animal has been estimated at up to fifty feet. There are several cases where witnesses saw what they thought was a floating log, only to have the "log" start to move and flex, showing humps or coils as it swam off.

One unusual feature of the Lake Okanagan case is the existence of several films and videotapes of the lake's residents in action. Some of these recordings, especially those showing only a "head," could be beavers or other small animals swimming. Those which are more intriguing, apparently showing a line of connected objects making coordinated movements, have all been shot at too great a distance to be definitive. At this writing, a videotape shot from a houseboat on August 9, 2004, is attracting a lot of scrutiny. Jon Manchester, news editor of the *Daily Courier* in the lakeside town of Kelowna, called himself a skeptic on the beast in general but said, "It's by far the most convincing Ogopogo evidence I've ever seen."

Those who think there is a strange animal involved in the Okanagan business have suggested all of what might, in the lake monster business, be called "the usual suspects," from the relatively mundane (a giant sturgeon) to the really exotic, like archaeocetes and plesiosaurs.

Okanagan often has excellent viewing weather, and the lake's water is clear. The lake attracts large numbers of boaters and vacationers. While a minority of visitors and residents have reported sightings, others have spent many seasons boating on the lake, or lived their entire lives on its shores, without noticing anything unusual. Interest increased when, in May 2000, a reward of two million Canadian dollars was posted for proof of the creature's existence. Sadly, the offer expired in September 2001 without an award.

The case of Lake Champlain, a body of water bordered by New York, Quebec, and Vermont, is somewhat similar. Champlain is also a large lake, over 110 miles long, twelve miles wide, and up to 400 feet deep. Once again, reports of a "horned serpent" in the lake allegedly go back to the native Iroquois. (In almost every case where a monster has been reported in a North American lake, someone has unearthed an "Indian tradition" to back it up. This may be due to the richness of Native American mythology, a genuine familiarity with strange beasts, or the oft-overlooked fact that Native Americans, like everyone else, have a sense of humor.)

It is commonly written that the lake's European discoverer,

Samuel de Champlain, reported seeing a monster in 1609. Where this belief came from is unclear. Champlain's record of his visit to the lake mentions no aquatic animals other than fish. In any event, there are over a hundred sightings from 1873 to the present of something generally resembling the monster reported in Lake Okanagan.

For example, vacationer Olivia Patten reported being thoroughly terrified by an encounter in 1981. She was swimming in the lake when a pair of thirty-foot-long "sea serpents" emerged near her. Not surprisingly, she had no interest in swimming in this particular lake ever again.

Dennis Hall, an archaeologist who directs a group of volunteer investigators called "Champ Quest," personally claims the two most unique sightings on record. According to Hall, he once saw a foot-long baby Champ. This had been captured by Hall's father. Years later, he allegedly watched an adult whacking the water with its long neck to stun fish, an activity which no known animal of any type has been observed to perform.

Joseph Zarzynski, the chief chronicler of "Champ," has written that about forty percent of the witnesses claim to have seen the animal's head and neck. As with Lake Okanagan, there are longtime residents of Lake Champlain who have seen nothing and doubt the phenomenon's existence. "It's a (expletive deleted) watersoaked log and all the people who see it are soaked too," Anthony Mydlarz offered.

There is also a famous photograph, taken by Sandra Mansi in 1977, which shows a plesiosaur-like shape with a swanlike neck and small head. Mansi believed the neck reached about six feet out of the water. Analysis of the picture has been hampered by two facts: the negative has been lost, and Mansi has been unable to find the exact spot from which she took the picture. Suggestions to explain the picture include a floating tree trunk, a waterbird, or an inflatable dummy "monster."

Mansi thought the object in the photograph was six feet high and 150 feet from shore. A test of an object that size at that distance indicated it should have appeared larger in the photograph, so Mansi apparently made an error in either the size or the distance. Neither is surprising when looking at an unfamiliar object on open water, but in which direction did the error occur? That is, did Mansi see a smaller object near to shore, or a larger one further away?

Those who believe the photo shows a real animal are divided on how to interpret this evidence. One of the favorite candidates for lake monsters, especially in Lake Okanagan, is an archaeocete. While these prehistoric whales were serpentine and could flex vertically, they could not possibly display the shape in the Mansi photograph. Neither could a huge eel. If this picture really shows Champ, one has to postulate either a plesiosaur, which has left no fossil record for sixty million years (and must have evolved a much more flexible neck) or something like a long-necked seal which has somehow left no fossil record at all.

To add to the identification problem, Lake Champlain freezes over completely in most years. Where would an airbreathing animal like a pinniped go? Could it hibernate, sheltered by caves or overhangs offering air pockets? Could it have evolved underwater breathing capability? Either is technically possible, but the farther out you have to go on an evolutionary limb to explain the animal's characteristics, the weaker that limb gets.

One non-monster explanation offered for the Champlain sightings concerns a phenomenon called a seiche, or standing wave. Researchers from nearby Middlebury College have found that, under weather conditions common on a long, narrow lake, a wave can build up that is invisible on the surface but thirty to sixty feet high underneath. This "sloshes" from one end of the lake to the other, sometimes for days. Such a wave could do strange things, such as making logs and other inanimate objects appear to move strongly against the wind.

Sightings in Lake Champlain continue to this day. The town of Port Henry has "adopted" Champ and uses the monster tales as a draw for tourists. The aforementioned Champ Quest organization mounts an ongoing effort to collect reports, conduct sonar searches, and otherwise investigate the thing.

Skeptics say there is nothing to investigate besides the forces of nature. Lake Okanagan, Lake Champlain, and Loch Ness are all long, narrow bodies of water. This makes them susceptible, not only to the seiche effect, but to wake phenomena. Boat wakes can be reflected from the banks, meet in the center of the lake, and throw up a line of waves long after the originating boat has passed from sight. Finally, when the surface of any large lake is still, conditions are often created for a type of mirage which elongates objects vertically, so a floating stick may become (briefly) a hefty "monster."

There are many other stories like those from Lakes Champlain and Okanagan. They originate in lakes around the world, from Ainslie (Nova Scotia) to Zeegrzynki (Poland). In no case has any form of physical evidence been collected. Nor are there any authenticated photographic records displaying the kind of clear, close-up detail needed to prove the existence of an unknown animal. If any of these animals exist, such evidence must eventually be found. If that evidence never surfaces, we will have to abandon the romantic idea that we share our lakes with creatures from a bygone era.

PROTECTING NESSIE

As the best known of all cryptozoology's "monsters," Nessie deserves the final word in this section. Cryptozoologists often feel put-upon and scorned, so a bit of humor is always welcome. Humor seems to crop up most often in connection with a certain beautiful and famous lake in Scotland.

In 1998, the British government released some official papers related to Loch Ness. These showed how, in 1967, officials had dealt with the question of whether to protect the Loch Ness animals. When leaders of a Japanese expedition announced plans to tranquilize the beast (one wonders how they planned to knock out a huge animal of unknown classification, but never mind), the local inhabitants queried the government about the legality of harming Nessie.

Finding the appropriate government department, however, was not easy. The Scottish Development Office ruled that, while laws required regulation and protection of freshwater fish, there was no proof Nessie was a fish, so it wasn't their problem. The question also went to the Department of Agriculture and Fisheries, which had no idea what to do with it, and so did nothing. One official wrote, "I think this letter has been in every department other than Monty Python's Ministry of Silly Walks—which might be the best place for it."

If you think Nessie is real, you can buy insurance against being harmed by it. Goodfellows, a London company which specializes in filling the unusual niches in insurance coverage, offers the Nessie-Safe All Risks Aquatic Insurance policy. For 260 pounds, you can be insured for a million pounds against death, disability, or post-traumatic stress disorder "occasioned by the Loch Ness Monster. "

No one has tried to collect.

SECTION IV: MISCELLANEA

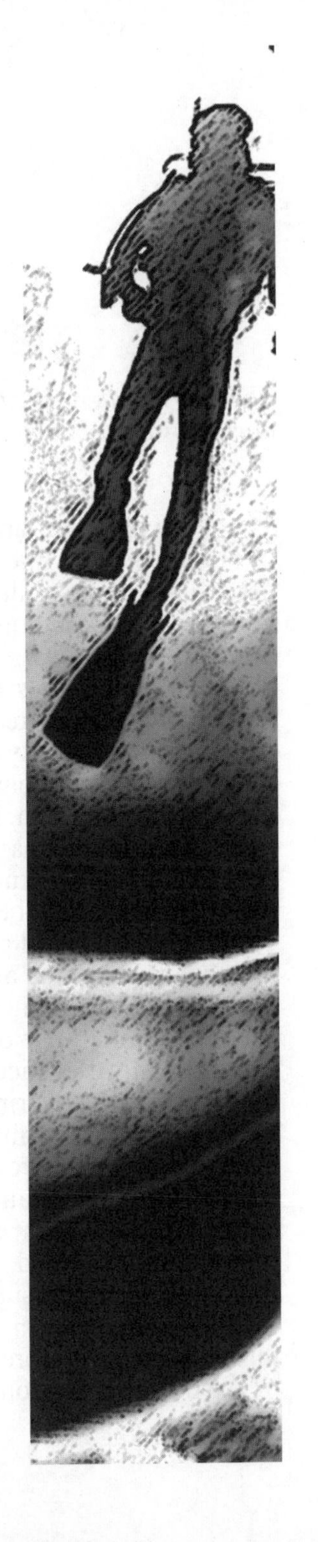

INTRODUCTION: SECTION IV

This world, after all our science and sciences, is still a miracle: wonderful, inscrutable, magical and more, to whoever will think of it.

— Thomas Carlyle

There are plenty of animal oddities, mysteries, and curiosities left to ponder. Some, like the identity of the seal-sized "sea monkey" naturalist Georg Wilhelm Steller reported in 1741, may never be laid to rest. Other questions, like the possible existence of the monstrous *Octopus giganteus*, will be resolved eventually, when we have better tools to search the vastness of the ocean depths. (After all, if there really is an octopus the size of a basketball court, it can't hide forever.)

Intriguing questions even surround such well-known groups as the North American land mammals. Europe, Asia, Africa, and Latin America all have their small, long-tailed wild cats. North America, as far we know, has none north of Mexico. It's not obvious why this should be so, and Pennsylvania cryptozoologist Chad Arment has gathered a body of old news items and letters suggesting that it may not always have been so. In the 1800s, several newspapers and hunters in Pennsylvania wrote about a true wild cat very unlike their domestic cats. It was larger than a domestic cat and had a flat face, prominent, sometimes tufted, ears, and a long tail which was ringed in black. The body was gray or brown, turning to buff or whitish on the underside, with a variety of black markings which were sometimes strong enough to give the cat a tiger-striped appearance. A folklorist named Henry Shoemaker collected numerous accounts of these animals being killed, trapped, or exhibited. The cat was always described as rare, and there have been no such reports in many years, And so a mystery animal right here in the United States just may have slipped through our grasp forever.

Also in this section are some solved mysteries. We know now there really wasn't a giant seabird wandering the beaches of

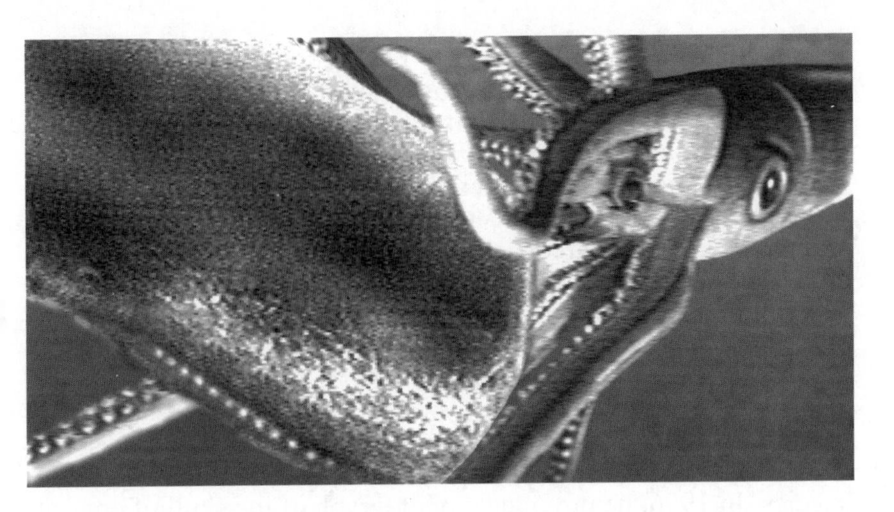

The Giant Squid, a seemingly impossible animal that turned out to be real.

Florida in 1948. We know the *ri*, a briefly-celebrated mystery creature from New Guinea, was just a dugong behaving strangely. And Oliver, a weird-looking chimplike primate once touted as a baby Bigfoot, turned out to be—well, a weird-looking chimp.

There are countless other puzzles there to keep the curious zoologists, naturalists, and lay enthusiasts of the world guessing. Have cetologists identified the elusive whale known as Mesoplodon "Species A?" How giant does the giant squid become? Do unusual cat skulls collected in Peru indicate the existence of unknown species? What is the truth behind such startling reports as the blue-furred tiger mutation reported from China?

Investigating such conundrums is an enjoyable, yet apparently endless voyage in search of knowledge. If most of the questions in this section turn out to have prosaic answers, they are still worth the attention of science. Only in searching for answers—whether those answers, in the end, delight us or disappoint us—do we learn about ourselves and our world.

PERUVIAN MYSTERY CATS

Peter Hocking, a Christian missionary who is also a naturalist affiliated with the Natural History Museum in Lima, Peru, has been chasing new mammals for years with the help of the Peruvian hunters he knows. He has heard of many mammals which might be new species, but the most intriguing are two possible new big cats.

Before we get to those, I should mention Hocking has at least one confirmed mammal discovery to his credit. In 1999, Hocking permitted me to announce in my newsletter *Exotic Zoology* the finding of a new coati in the genus Nasuella. Hocking received the first (dead) specimen from a farmer in 1998. The animal, he was told, lived in cloud forests in the Peruvian state of Apurimac. In 1999, he found two more specimens at a local zoo.

Hocking is working with Victor Pacheco, a mammologist at the University of San Marcos in Lima. Pacheco confirms the discovery is a new species or very distinct subspecies and is preparing the description. The only known species in the genus Nasuella lives only in Columbia and differs in color. The common coatis of the genus Nasua are more gracile and longer-legged, with longer, thinner tails.

Of potentially greater significance, Hocking has obtained skulls of what he believes are two big cats, one of which may be a new species. In 1996, he provided a photograph of the skull of one cat, the "speckled tiger," sitting next to the skull of a jaguar (*Panthera onca*). This "tiger" is reportedly gray with black spots.

I sent the photograph to two mammologists, Dr. Troy Best of Auburn University and Dr. Cheri Jones of the Denver Museum of Nature and Science. According to Dr. Best, the photograph by itself was simply not sufficient evidence by which to make a judgment. Dr. Jones noted cautiously that one skull appeared to be a jaguar's and the other "seems to be of a larger felid."

In addition to being larger, the "tiger" skull has more prominent canine teeth than the jaguar skull and a different ratio of width to height. (The width-to-height ratio of the jaguar skull is approximately 1.6 vs. 1.4 for the tiger skull.) None of this rules out the tiger skull's coming from a jaguar—only expert comparison with a large number of jaguar specimens could do that conclusively—but it's intriguing.

Hocking obtained a skull of another reported cat, which he calls the "striped tiger." In 1997, I attended a meeting at the Denver Museum of Natural History where Hocking reviewed photographs of both skulls with Dr. Jones. She agreed further investigation was warranted and referred Hocking to two other mammologists, both felid experts. Hocking suggested the "speckled tiger" may actually a be a jaguar, but a previously unknown color morph, while the "striped tiger" (which is reportedly rufous in color with white vertical stripes) was more likely to be a new species. If this proves true, Hocking could have the first new big cat species described in nearly a century and a half. The hunter who obtained the skull of this latter cat sold the skin, which Hocking has attempted unsuccessfully to track down and repurchase. At this writing, Hocking's cat skulls are still awaiting proper study.

New big cats are always approached with caution, both literally and taxonomically. Big cats are the top predators everywhere they exist, so there will never be many species in the same habitat. Nowhere on Earth today do more than three such species overlap. More commonly, there are only one or two species: the puma essentially has the huge range of the United States and Canada to itself. Most of South America is home to both the known jaguar and the puma. In the Pleistocene Era, the continent had at least two more big cats, but it also had more ungulates and other large prey species.

Dr. Louise Emmons, who as we saw in Section I also has experience finding new mammals in Peru, was highly skeptical when asked about Hocking's cats. She pointed out there is no vacant ecological niche available for another big cat to occupy. That doesn't mean the existence of such a cat—perhaps a relict population with a restricted range—is impossible, but it does mean it's unlikely. (The same logic, she thought, applied to Marc van Roosmalen's black jaguar.)

Peter Hocking believes that, regardless of the odds, there are indeed unknown big cats in Peru. Given the remoteness of much of the land involved and the elusiveness of felids in general, it may be a long time before he can prove that belief.

QUEST FOR THE GOLDEN BEAR

The eight known species of bears constitute another line of mammals with which humans have a special fascination. Bears range from the impossibly cuddly (the panda) to the downright terrifying (the great brown bear, whose subspecies are known as the Kodiak, grizzly, etc.) The polar bear has adapted to a semi-maritime existence north of the Arctic Circle, while small sloth and sun bears saunter through the heat of tropical jungles.

Can it be that some species of these great carnivores remain undiscovered? It seems unlikely, but there are some intriguing mysteries and oddities in the bear world.

In the Smithsonian rests a buff-colored pelt and an odd skull, taken from a 600-lb bear killed in Canada's Northwest Territories in 1864. Mammologist C. Hart Merriam thought the skull and teeth were so odd that the bear must be some sort of relic of an ancient lineage. He proposed a new genus and species and named it *Vetularctos inopinatus*. Modern experts are undecided on whether this could have been the last of a forgotten species, a grizzly-polar bear cross (something known in zoos, though not confirmed in the wild), or an unusual-looking grizzly which fell victim of Merriam's penchant for creating new species based on minor differences. The evidence is still in storage, waiting for the modern mammologist with the time and interest to take a new look.

Across the Bering Strait, there are also odd bear tales. In the 1920s, zoologist Sten Bergman named a new subspecies of the brown bear, *Ursus arctos piscator*, based on a huge bear skull and a pelt with short black fur which dwarfed any bear he had ever seen. *U. a. piscator*, if it was valid, is usually considered extinct, although reports of black giant bears lingered at least into the 1980s.

More recently, naturalist/writer Sy Montgomery and biologist Gary Galbreath followed tantalizing hints of another bear all the way to Southeast Asia. This region is host to two known bears, the moon bear, or Asiatic black bear (*Selenarctos thibetanus*) and the smaller sun bear (*Helarctos malayanus*). Both species are normally dark with a light-colored chest blaze, though several variations like a brown-phase moon bear occur. Immigrants from Laos to the United States told Montgomery and Galbreath of another animal, the "dog bear." Dr. Galbreath had already seen, in a village in China, an unidentified animal which resembled a moon bear, but with a stunning golden coat. The threads of these bear tales intertwined and crossed with other hints and riddles to lead the two Americans on four expeditions beginning in 1999.

Montgomery's enthralling book, *Search for the Golden Moon Bear*, describes the resulting quest. Centering their efforts on Laos but also inquiring in Thailand and Cambodia, Montgomery and Galbreath talked to government officials, tribal hunters, and old villagers, hearing a welter of conflicting tales about golden bears, huge black bears, and even stranger varieties. Along the way, they found caged bears of all types, including moon bears with golden coats ("blond" is the proper word in biology, but it really does not do these animals justice). They also saw heartbreaking massacres of wildlife, a society in disarray as knowledge of nature disappeared along with traditional cultures, and an apartment museum stuffed with thousands of pairs of antlers from creatures common, rare, and extinct, not to mention a python skin forty-five feet long. (Snake skins can be stretched quite a bit after death, but this must still have been one monster of a snake, perhaps a world record.) The adventurers looked along the way for evidence of the mysterious khting vor, although they found nothing to reinforce the reality of the animal.

In the end, Montgomery and Galbreath did not find a new species. They did find several examples of the golden bear, showing several gradations in color, and documented these for the first time in a scientific paper. Most such bears had a mane, almost lion-like in appearance (although shorter than a lion's) and darker than the rest of the animal's coat. The travelers even observed a never-reported "panda" variety with darker forequarters and dark rings around its eyes. Their work also extended the known range of the moon bear—tragically, just in time to watch the Laotian population hurtling toward extinction.

Galbreath noted that the question of species remained a bit of a mystery. Over the last century, writers, hunters, and other witnesses have contributed enough accounts of unusual or seemingly out of place bears that it remains just possible the jumble of facts and tales does conceal another species. Montgomery wrote that she, too, thought this was possible. Her hope was that, if there was such a bear, it might survive long enough for scientists to find it.

REQUIEM FOR THE GIANT OCTOPUS?

It is an odd fact that the largest, strangest mystery animal in the cryptozoo is also one of the few whose existence is backed up by photographs and tissue samples as well as eyewitness accounts.

This story began when a mass of whitish organic matter weighing an estimated five tons drifted onto the beach at St. Augustine, Florida, in 1896. The main lump was tough and rubbery and measured over twenty feet in its longest dimension.

DeWitt Webb, a medical doctor who headed the local natural history society, believed the carcass was a giant octopus. He sent photographs and tissue samples to Dr. A. E. Verrill, the leading cephalopod expert of the day. Dr. Verrill at first thought the thing was a giant squid. He later came to agree with Webb on an octopus, which he labeled *Octopus giganteus*. Later still, he decided it was just part of a whale.

This puzzled Webb, who wrote, "It is simply a great big bag and I do not see how it could be part of any whale." Descriptions written by Webb and others mentioned tentacles, a "mantle," and a single large baglike internal organ, so the mass was not entirely featureless. No account of the incident mentions bones, cartilage, teeth, baleen, fins, or flippers.

There have long been reports, mainly by fishermen, of a giant octopus in this region. While we are on the topic of the giant octopus, I should clear up a bit of history. In *Rumors of Existence*, I cited a story reported in Gerald Wood's book *Animal Facts* and Feats. This concerned a U.S. Navy ship, a fleet oiler named the USS *Chicopee*. According to Wood, in the spring of 1941, two

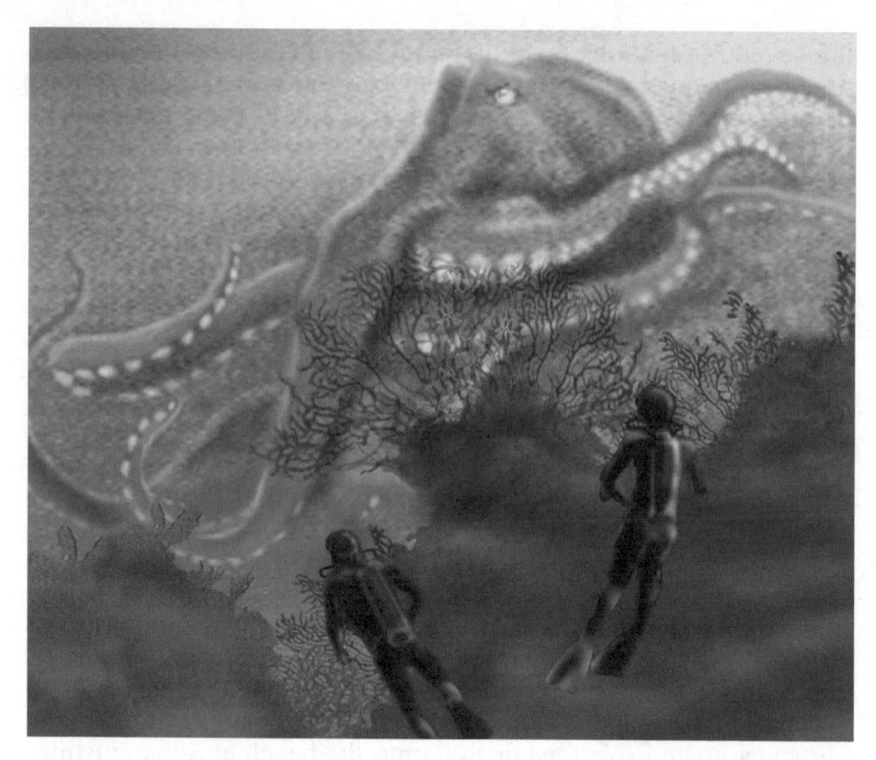

One artist's rendition of a giant octopus.

destroyers dropped depth charges off Fort Lauderdale, Florida. (He did not say why live depth charges were dropped, but this could have been an exercise or part of the undeclared but already real war against German submarines.) Sailing close behind came the *Chicopee*, whose crew investigated what looked like debris and discovered a titanic brown octopus, floating in an apparently stunned condition.

In checking further, though, I found the *Chicopee* was not commissioned until February 1942. I was able to contact Art Bieberstein, formerly a radioman on the *Chicopee* who maintained one of the ship's logs (a U.S. warship has several logs maintained by various departments) and had kept a personal copy. He was on the ship from September 1943 through the end of the war and was certain that nothing so memorable took place. It could have happened between the ship's commissioning and the date Bieberstein came aboard, but it's hard to imagine that newcomers to the vessel would not be regaled with stories of such a dramatic incident. It appears that either the octopus story is a hoax, source unknown, or the events have been attributed to the wrong ship.

The St. Augustine lump eventually washed out to sea, and the subject was forgotten until two scientists, Forrest Wood and Joseph Gennaro, retrieved a tissue sample from the "mantle." The specimen had been sent by Webb to the Smithsonian and forgotten about until these two researchers dug it out. They wrote in 1971 that, based on microscopic examination of the tissue structure, the thing was an octopus, with a total armspan approaching 200 feet.

Dr. Roy Mackal, a biochemist and cryptozoologist, in 1986 published his analysis of the chemical composition and amino acids present in the tissue. He wrote, "It was essentially a huge mass of collagenous protein. Certainly, the tissue was not blubber." He agreed with the octopus identification.

Richard Ellis, a skeptic of most sea-monster claims, stated the same conclusion in his 1994 book *Monsters of the Sea*, which included a photograph showing what appeared to be tentacles. Ellis opined, "If one examined the pictures with no information as to the size of the animal, there would be no question ... It was an octopus."

A quartet of marine scientists led by Dr. Sidney Pierce and including Dr. Eugenie Clark (who, as a board member of the International Society for Cryptozoology, could not be called closed-minded) tested the tissue again in 1994. They also examined a mass which washed ashore in Bermuda in 1988. They reported that neither tissue structure nor chemical composition indicated an octopus. The Saint Augustine carcass was probably whale skin, while the "Bermuda Blob" came from a shark. Pierce and his colleagues suggested the whale's fatty tissue had decomposed, leaving only the collagen "network" that runs through whale blubber.

All the pro- and anti-octopus researchers agree on three points: 1) the mass as beached showed no evidence of any type of skeleton, 2) the samples were almost pure collagen (connective tissue), and 3) the carcass was not a giant squid.

Mackal, in his analysis of amino acids, considered proline and glycine levels to be most likely unaffected by age and preservatives. Levels of both were much higher in *O. giganteus* than in the dolphins Mackal used for comparison, but also somewhat higher than in known octopus and squid. He did not do a comparison to a large whale. Pierce's team found *O. giganteus'* values closer to those of a whale than an octopus, but, while Mackal reduced all his samples himself at the same time, the 1994 analysis used comparison values for a whale of unknown species analyzed by another researcher in 1955.

The 1994 analysis did not explain the physical characteristics of the St. Augustine mass, and Richard Ellis was not the only investigator who asked how the skin of a whale could become detached from its owner. Whalers peeled blubber off in long strips: orcas may rip off chunks, but nothing like the mass involved here. Accordingly, Ellis wrote, the four scientists had forced "a cryptozoological creature into a recognizable category, even if the fit is a very poor one ... it is difficult, if not impossible, to explain how the skin of a whale could become a five-ton blob of collagen."

The debate flared up anew in the spring of 2003, when a gray, featureless blob forty-one feet across and nineteen feet wide drifted onto Pinuno Beach in Chile. Elsa Cabrera, a biologist with the Center for Cetacean Conservation in Santiago, announced the remains were not from a whale and could belong to a giant octopus. Experts in other nations, examining the photographs, disagreed. Most suggested it was indeed a whale. Cryptozoologists protested that it was hard to believe a scientist with a cetacean institute could misidentify a dead whale when she was literally standing on top of it.

It turned out, though, that such a misidentification was possible. The Chilean blob was positively identified as the remains of a sperm whale. The whale apparently floated long enough that the skeleton and most of the internal organs fell away, producing a mass with no obvious cetacean characteristics.

The Chilean episode obviously strengthens the possibility that the 1896 carcass was indeed a decaying cetacean. While there undoubtedly remain many unknown cephalopods creeping on the ocean floor, and fishermen still make strange reports from the Bahamas, *Octopus giganteus* is likely to be forgotten until—and unless—it makes another appearance in the flesh.

MYSTERIES FROM VU QUANG

As we've seen, a host of fascinating new mammals has already emerged from the shadowy jungles of Southeast Asia. Are there more to be found? It would seem so.

During his journeys of discovery in the region, the WWF's Dr. MacKinnon found evidence of two more large bovids, still unclassified. Vietnamese biologist Ngoc Chinh showed him the skull of the *quang khem* (slow-running deer). This came form a region called Pu Mat, just north of Vu Quang. MacKinnon found more skulls of the same animal, overlooked for decades, in the collection of a Hanoi scientific institute. The skulls show the animal has simple spikes for horns, and its DNA does not match any known species. Finally, MacKinnon found a pair of antlers in the same collection which didn't match anything discovered so far, including the quang khem. These, he was told, belonged to the *mangden*, or black deer.

Finally, there is the unknown bovid described in a *South China Morning Post* story dated January 7, 1995. This stated that two animals were captured near the village of A Loui in central Vietnam. The adult specimen escaped, but its calf died and was eaten. This mammal had long ears, a round head, and a stout body covered in black and gray fur. Vietnamese biologist Ha Dinh Duc was quoted as saying the animal appeared to be a new species.

These three species have still not been tracked down and identified, but they're out there. The greatest mammal "gold rush" of modern times is not over.

FISHING THE OCEANS

Of the estimated 50,000 vertebrates believed yet to be classified, most are saltwater fishes. Sometimes we have had tantalizing glimpses of species which remain unknown to science.

Two researchers who manned the submersible *Deepstar 4000* on a 1966 probe of the eastern Pacific, for example, had an uncomfortably close encounter with an awesome denizen of the deep. They were motoring along at a depth of 4,000 feet in the San Diego Trough when a dark-colored, mottled fish estimated to be thirty to forty feet long swam right up to the eighteen-foot sub. The fish studied the craft with eyes "as big as dinner plates," then moved off, much to the relief of the startled aquanauts.

Automatic cameras lowered into the same area took pictures of a large fish identified as a rare Pacific sleeper shark. If this was what the Deepstar met, it would be, by far, the largest sleeper shark ever seen.

The witnesses, pilot Joe Thompson and oceanographer Dr. Eugene LaFond, doubted their visitor was a shark. Both men described a round tail like a grouper's rather than a sharklike tail. Additionally, the eyes described were much too large for a sleeper shark. So it's reasonable to conclude this incident involved a new and gigantic species. Was it, as the description indicates, a bony fish or teleost? If so, it would be the largest member of the entire class Osteichthyes, easily outmuscling the current record-holders for length (a title given by Guinness to the oarfish (*Regalecus glesne*) at twenty-five feet) and weight (the bizarre ocean sunfish (*Mola mola*) at nearly 5,000 pounds.)

There are many other mysterious deep-water fish, especially sharks. One odd-looking type, six feet long with bulging eyes, was reported from a French bathyscaph off western Africa at 13,000 feet. A 1986 reference listed fourteen species of shark known from a single specimen each, and the difference between one specimen and none is mere chance. One example was the Antarctic sleeper shark, described in 1939 from a single eight-foot carcass which washed up on Macquarie Island in 1912.

No fewer than forty-two new species of sharks were described from 1985 to 1995. At that point, there were at least forty-five potential new species awaiting confirmation and classification. This means eighty-seven species of sharks were either discovered or possibly discovered in one decade.

What else may be waiting for us in the depths? Mac McCamis, a pilot of the submersible *Alvin*, once reported a glimpse of something he didn't want to see any closer. He called it "this monster ... kind of shook me up. This was a living creature ... I seen at least forty or fifty foot of it." Not a very helpful description, but certainly a tantalizing one.

Someday, when we have the technology to thoroughly explore the deep oceans, we will find these creatures ... and many more.

SURVIVING SLOTH?

Is some type of huge, prehistoric ground sloth still living in Brazil? A Brazilian ornithologist, David Oren, believes the creature known as *mapinguari* to local hunters and farmers may well be real and may represent just such a survivor.

Oren is the head of zoology at Emilo Goeldi Natural History Museum. In the course of his work collecting and observing animals, he heard many tales of the mapinguari. While these stories are thought by some cryptozoologists to refer to a primate, somewhat like sasquatch, Oren was struck by the way descriptions of the red-haired beast corresponded with what we know of the mylodontid ground sloths. There were several species of ground sloths in the Americas. The largest, *Eremotherium laurillardi*, reached the size of a respectable elephant at around three tons. Some South American species were definitely hunted by early humans and are believed to have been alive only 8,500 years ago.

Features of the mapinguari reports which match some ground sloths include general size and appearance, tracks showing clawed toes pointed inward, and the color (samples of ground sloth skin have been found preserved in caves).

According to the people of western Amazonia, the mapinguari is a shaggy animal weighing about 600 pounds. This would make it one of the smaller ground sloths, but it would still be the largest living mammal native to South America. Oren believes the mapinguari is nocturnal and vegetarian. It rears up to a height approaching six feet when startled and is supposedly accompanied by a stench which could gag a jaguar. Whether the animal is still living or has gone extinct in the last few years is a question Oren cannot answer, but proving either possibility would be a scientific coup.

In 1995, Oren launched his first expedition into the Brazilian rain forest in search of the mapinguari. This seems an outlandish pursuit for a man with a Harvard doctorate and a solid scientific reputation. Oren, though, finds the eyewitness accounts and eleven-inch tracks compelling. He also knows how much publicity would accompany the discovery of such a spectacular animal—publicity that would lead to increased protection for the hard-pressed rain forest. So he has pressed on, persevering through six arduous expeditions and a growing thicket of skepticism from his colleagues. He has also attracted some assistants. Writer Marcelo Volcato, in 1999, located a new witness in the state of Matto Grosso who described a creature with long reddish hair, standing over five feet tall on its hind legs and smelling very bad. The man was familiar with regular sloths and saw no resemblance, although his description resembles nothing else known in Brazil either.

Professor Paul Martin, an expert on megafaunal extinctions, thinks Oren is wasting his time. "I'll eat my share of sloth dung if this animal is alive," he declared. While Oren has said, "I'll be the first one to admit the whole idea is rather absurd," he also argues, "I'm testing a scientific hypothesis that's basically reasonable. This isn't the Loch Ness monster."

Oren's work so far has ended in failure: hair and fecal samples he's gathered have been identified as those of known animals. But Oren is steadfast. One of his sympathetic colleagues, Phil Hazelton, points to the Chacoan peccary discovered in Peru in 1973. "That," Hazelton notes, "was a mythical animal as well."

A WEIRD FISH—EVEN BY CALIFORNIA STANDARDS

In 1993, the *San Francisco Examiner* published an item describing a bizarre fish caught in California's Clear Lake. About thirty inches long, the critter resembled a catfish, but it had a strangely shaped head, thick, fleshy-looking fins, and—weirdest of all—a horizontal tail, something possessed by no known fish, living or extinct.

Curious, I contacted Richard Moreno, a scientist with California's Fish and Game department. He reported the creature was finally identified as a bizarrely mutated channel catfish.

Moreno agreed it is surprising such an extreme mutation survived to adulthood. He noted the catfish in Clear Lake are not native, but come from hatcheries, where mutations are more common than in the wild. And so it was that a potentially stunning discovery was reduced to an interesting footnote.

FOSSILS FROM ANCIENT FORESTS

Botany, even cryptobotany, may seem to lie outside the subject matter of this book, but two recent events from Australia have encouraging implications for cryptozoologists.

In 1994, ranger David Noble discovered a stand of thirty-nine strange-looking pine trees. These grew in a tiny, steep-sided valley in the Blue Mountains, part of Wollemi National Park. Noble hiked a long way to reach the remote valley, 125 miles from the city of Sydney, and may have been the first human ever to set foot there. These trees completely puzzled him, and he carried a sample out to show to botanists.

The botanists were stunned. Noble's pine trees had supposedly been extinct for fifty million years. Carrick Chambers, director of Sydney's Royal Botanic Gardens, marveled the tree discovery was "the equivalent of finding a small dinosaur still alive." The trees, according to one description, have "dense, waxy foliage and knobby bark that makes them look as if they were covered in bubbly brown chocolate." The tallest of this previously unknown species of the family Araucariaceae stands 115 feet high. Isolated valley or not, this makes *Wollemia nobilis* (as the species was eventually named) a very large life form to go undiscovered. A second stand has since been found in another valley.

No sooner had botanists digested the discovery of this forest fossil than a tree from northeastern Queensland was identified by Dr. Andrew Douglas as the only known survivor of a family of flowering trees, the Proteaceae. This type had also been presumed extinct for tens of millions of years.

Given the central role trees play in their ecosystems, it's reasonable to predict some new insects and other small creatures living on or around both of these new species will also be found. It's also reasonable to expect Australian botanists will look around very carefully when they're in the wilderness.

PRIMATES IN THE SHADOWS

Since we humans are primates ourselves, we have a natural fascination with our cousins, the monkeys and apes. There have been at least fifteen species of monkeys and other small primates discovered in the last decade. The question asked in this essay is, "What other species do we have evidence for?"

I'm not talking here about such controversial subjects as sasquatch and the yeti. Instead, this is an effort to survey the discoveries, reports, and mysteries concerning more "normal" primates.

Let's begin with South America, where new marmosets and other small primates have been appearing with startling frequency for the last several years. As described earlier in this book, there have been more than a dozen from Brazil alone, plus others so new their descriptions have yet to be published.

There are scattered but persistent reports of larger primates in several areas of the continent. The most famous of these claims is almost certainly a hoax. This is Swiss geologist Francois de Loys' 1920 photograph of an alleged Venezuelan ape. This claim initially gained some acceptance, and a French anthropologist christened the animal *Ameranthropoides loysi*.

Today, however, De Loys' ape has almost universally been rejected by primatologists. It is very difficult to accept that an unknown New World ape has evolved to look almost exactly like a spider monkey from the genus Ateles. It does look like a strikingly large and robust spider monkey, raising the possibility De Loys found a new monkey subspecies or species but couldn't resist embellishing the discovery.

That said, there are stories of large primates of varying descriptions (some upright and tailless, others clearly monkeylike) from several countries in Central and South America. Peter Hocking has collected many reports from Peru of the *isnachi*, most often described as a large, robust black monkey with a short tail. Another example is a five-foot-tall, bipedal primate reported from Guyana in 1987. Dr. Gary Samuels, an American mycologist, says he had a very clear look at this creature.

If any of these reports are accurate, they could describe a genuine New World ape (which would be as big a shock to zoologists as finding the Loch Ness monster) or a very large monkey. The monkey would be almost as surprising, but not impossible, especially given the recent discovery of a suitably robust fossil species named *Protopithecus brasilensis*. Needless to say, more evidence is required.

British conservationist Debbie Martyr, aided by wildlife photographer Jeremy Holden, has pursued Sumatra's *orang-pendek* ("short man"), also called the *sedapa*, since 1989. She has collected footprint casts and numerous local accounts of what is reportedly a four-foot-tall, habitually bipedal primate. Both researchers also claim personal sightings. If the orang-pendek is a unique animal, it may be a distinctive subspecies of a known ape. Orang-utans, siamangs, and white-handed gibbons live on Sumatra. This marks the island as good ape habitat, but also offers several candidates for cases of mistaken identity (although gibbons and siamangs are smaller than the reported orang-pendek). Martyr believes the orang-pendek's habitual bipedalism and humanlike footprints establish it as something different. Hair samples found by a British expedition in 2001 have not been definitely identified.

The search for the orang-pendek, while not concluded, is quite "respectable" as cryptozoological quests go. No less an authority than John MacKinnon once came upon what he believed were sedapa tracks, which looked like a human child's prints but with a more pointed heel. On the subject of how he and others have missed seeing the live animal, he commented, "Sumatran rhinos are big, rare, and stupid, and I've never been able to see one despite spending years roaming around forests where they live…It would be a lot more difficult to see a sedapa, which is smaller, smarter, and just as rare."

When the discovery of Flores Man was announced, Dr. Henry Gee of *Nature* wrote, "In the light of the Flores skeleton, a recent initiative to scour central Sumatra for 'orang pendek' can be viewed in a more serious light. This small, hairy, manlike creature has hitherto been known only from Malay folklore, a debatable strand of hair and a footprint. Now, cryptozoology, the study of such fabulous creatures, can come in from the cold."

Martyr, at this writing, is working in Sumatra's Kerenci Seblat National Park. There she looks for primates when time allows while pursuing her most urgent work, the conservation of the highly endangered Sumatran tiger. Martyr is certain she has observed an unclassified ape (probably closest to the orang-utan *Pongo pygmaeus*, but either a separate a species or highly divergent subspecies). Tragically, she fears, the orang-pendek may vanish before it is discovered. She wrote in December 2003 (before her work was indefinitely disrupted by the tsunami disaster) that, "since most of our old study areas are now either heaving with illegal loggers or have been turned into palm oil, it is difficult to be very optimistic any more about this animal's future."

The most recently described ape came from Africa. The bonobo, or pygmy chimpanzee (*Pan paniscus*), was known and even held in captivity for many years before primatologists realized in 1929 it was distinct from the "standard" chimp (*Pan troglodytes*). There are still some mysteries about chimpanzees, and the number of species may not stop at two.

The most celebrated chimp in cryptozoology is Oliver, a former show attraction now living out his days at a primate retirement center. While Oliver is a strange-looking creature, habitually bipedal and with a high, bald-domed head, widely published claims that he had too many chromosomes for a chimpanzee have been disproved. He appears to be just a very odd chimp.

An aura of uncertainty continues to envelop another unusual-looking chimpanzee, the *koolookamba* (or *koolakamba*). This ape was so named by Native hunters in Gabon because of its strange "kooloo" call. The first specimen was collected by Paul du Chaillu in 1858. Since then, several examples have lived in captivity, and two were alive in the United States in the 1980s in a colony run by the Coulston Foundation. One of these, Minnie, was noted for her frequent bipedalism, unusual intelligence, and assertiveness as well as her appearance.

Koolookambas are large for chimps, with ebony-black faces, heavy brow ridges, powerful jaws, and wide, flat noses. They are alleged to live singly or in smaller groups than other chimps. Du Chaillu thought, and some zoologists initially agreed, they were a separate species of chimpanzee. This theory has been largely abandoned. Another idea is that the rather gorilla-like facial features of koolookambas might indicate they are chimp-gorilla hybrids. No recorded attempts have been made to mate chimps and gorillas to determine whether this is possible, although some apes do hybridize. (For example, siamangs and gibbons produce viable "siabons.")

Some alleged "kooloo" specimens have been identified definitively as small gorillas. This raises another complication, because such diminutive gorilla specimens have occasionally been assigned to a strongly disputed "pygmy" gorilla species, *Gorilla mayema*.

In 1964, Professor W. C. Osman-Hill classified the koolookamba as one of four subspecies of the chimpanzee, *Pan troglodytes*. Hill's identification is also controversial. While kooloos appear most often in mountain habitats, they apparently crop up in unrelated groups in a variety of locations. This indicates they may be just a variety, like the strikingly marked "king" cheetahs. Another anthropologist, Brian Shea, reviewed the problem in a 1984 article and suggested that, while there might be a distinct animal involved, all koolookamba specimens were most probably misidentifications of either large male chimps or small female gorillas.

Dr. Karl Shuker is not sure the mystery has been laid to rest. In a 1996 article, he wrote that the kooloo's resemblance to the mountain gorilla and preference for high altitudes might mark it as a race or emerging race deserving special study. Other researchers have made the same suggestion, and there may yet be unanswered questions concerning these curious apes.

Then there is Amman's "Bili ape." Given the size and other characteristics of the chimps in this population, it's interesting to speculate on a link between the Bili ape and the koolokamba tales. The Bili apes are not merely unusually large (the weight of an adult Bili ape has been estimated at 180 to 225 lbs., well above the maximum of 155 previously recorded for chimps), but visually distinctive, with flat, black, rather gorilla-like faces and a tendency to turn gray all over as adults.

In 1997, anthropologists announced the possible discovery of yet another chimp subspecies. Professor Don Melnick and his colleagues tested DNA from chimps living in various locations and concluded the apes of Nigeria and western Cameroon are as different from all other chimps as the known subspecies are from each other. Since the delineation of any subspecies usually involves a judgment call, Professor Melnick put the options this way: "You either collapse the other subspecies (into fewer, more variable ones) or name a new one."

If we could miss puzzling primates like the Bili apes until so recently, how sure can we be about other denizens of the world's remaining mountains, rain forests, and savannahs?

THE MYSTERIES OF WHALES

No one can fail to be captivated when looking into the eye of a great whale, even if it's only on film. The overwhelming impression a human observer receives is one of age—of being in the presence of an ancient intelligence possessing long-held wisdom and deep secrets.

It is, of course, only an illusion. Perhaps, though, it is an illusion concealing a grain of truth.

The strange truth about whale longevity surfaced only recently. Alaskan biologists examining bowhead whales (*Balaena mysticetus*) killed legally by subsistence hunters in the 1980s were puzzled to find stone and ivory harpoon heads at least a hundred years old embedded in the blubber. Tissue samples were sent to Jeffrey Bada and his colleagues at the Scripps Institution of Oceanography. There they were dated using a technique based on measuring amino acid levels, which change as a whale ages. In 2001, the results were published. If the analysis was accurate, several of the whales examined were well over a century old, and one had lived no less than 211 years. When this particular animal had been a calf exploring the Arctic waters beside its mother, George III was on the throne of England and the United States did not exist.

The bowhead is a strange animal to begin with. It has a blunted, arched head that looks like the result of a high-speed collision with an iceberg. It has the most northerly range of any whale, and it cruises the Arctic latitudes protected by a two-foot layer of blubber coating its sixty-ton body. As impressive as the species may be, no one until recently had an inkling it or any mammal lived to such an age. (The maximum age of blue whales has been estimated at over 110 years, and a very few humans have reached 120.)

The issue of longevity aside, there is no doubting that the order Cetacea does indeed hold questions human science has yet to answer.

The beaked whales in particular have long been an enigma to cetologists. A 1987 article in *Marine Mammal Science* noted twenty-four "positive sightings" of an unclassified species, over fifteen feet long, in the eastern Pacific Ocean. The adult male of this mystery species sports a distinctive creamy striping pattern down the sides of its otherwise a black or dark gray body, with the broad stripes coming together behind the head in a chevron. The females and young are reportedly brown on the back and lighter gray on the underside. The animal has been seen in pods of as many as eight individuals. This cetacean has long been referred to as the Unidentified Beaked Whale or Mesoplodon Species "A."

For many years, a logical suggestion was that Species A might represent the Indopacific beaked whale, *Indopacetus pacificus* (or *Mesoplodon pacificus*—opinions on its classification vary). This theory made sense when *I. pacificus* was known only from two skulls washed up thousands of miles apart. Now that we finally know what the Indopacific beaked whale looks like, we know it is not the mystery whale.

In 2001, marine ecologist Robert Pitman published his theory that Species A may be the adult form of the recently described pygmy beaked whale, *Mesoplodon peruvianus*. Despite Pitman's status as an expert on the mesoplodonts, there are problems with this identification. The specimens of *M. peruvianus* collected so far are smaller than the maximum size reported for Species A. Also, for the *M. peruvianus* identity to be correct, the telltale white bands on the male of Species A must fade considerably when the animal dies, as none of the stranded specimens of the former show more than faint markings.

Finally, these theories could all be wrong, and Species A may be a totally unknown cetacean.

If so, it is probably not the only unknown species. Pitman has twice spotted a Mesoplodon-type whale he calls "Species B." This is a dark gray animal with a strikingly long snout and a pale spot behind the eye. Pitman recently commented concerning beaked whales, "I don't think we have accounted for them all." Mere Dalebout agreed, writing that, "There are indeed still some sightings around of beasts which do not appear to correspond to any known species."

In 1995 and again in 1996, wildlife writer Jeremy Wade filmed a bizarre cetacean in the inland waters of Brazil. The animal looked like the Amazon river dolphin (*Inia geoffrensis*) but had a saw-toothed ridge down its back instead of a dorsal fin. When the animal arched its back at the surface, Wade thought it looked like a giant circular saw or "a gear wheel in the river's workings." According to local fishermen Wade questioned, this animal or one like it has been seen in the same stretch of river for over twenty years. This is either the strangest species in the order Cetacea or, more likely, the most extreme individual anomaly.

There are several observations on record of a cetacean about the size of an orca, but with a strikingly different dorsal fin. The orca has a magnificent sail-like fin which may be six feet high. The mystery whale has an equally high fin, but one so narrow in profile that observers compare it to a sword. This whale was described at close range by E. W. H. Holdsworth, a Fellow of the Zoological Society, in 1872. Holdsworth made his observations off the coast of Ceylon (now Sri Lanka) in 1868. The back of the whale was a solid grayish black, without the gray "saddle" and white head markings of orcas. Holdsworth's Ceylonese crew members had seen the whale before and called it the "Palmyra fish."

Sir James Ross reported "a line of large whales" with "remarkably long, pointed, black fins" in an 1841 account nearer to Antarctica. While Ross said "large whales," no truly large whale has a dorsal fin resembling what he was describing. Either "large" was a relative term and these were the same type of whales Holdsworth saw, or an entirely different species was involved. Other Antarctic explorers have seen whales with similar fins. This cetacean was sketched in 1902 by a prominent naturalist, Edward A. Wilson, who reported observing a group of four such whales.

All the observers were certain they were not seeing orcas, which are about the right size but whose striking color pattern makes them instantly identifiable. Whales fitting this general description were also reported in 1910 and 1911. In 1964, a scientific expedition led by Dr. Robert Clarke spotted and photographed an unidentified species of whale off Chile on eight occasions. The whales, traveling in groups of fourteen or more individuals, were compared by Clarke and his colleagues to Wilson's whale. The size and color matched, as did the geographical area. These whales were up to twenty feet long and very dark, although there were gray and white streaks on the forebody. The dorsal fin was not as high as Wilson reported, but Clarke still felt the two incidents may well have involved the same species.

A puzzling observation recorded in 1993 might also concern this mystery species, although the description is not a perfect fit. The officers of the ship *BP Admiral* were in the north Pacific on June 12 when two unidentified whales were sighted. The larger whale had a visible length over thirty feet, while the other showed about ten feet of its length and may have been the first whale's calf. Both whales were dark gray or black on top and cream-colored or white underneath. They did not show the distinctive coloration of orcas, but they did display striking, sickle-shaped dorsal fins.

The fin on the adult whale looked very much like those in Clarke's photographs. Again, it was not as high as on Wilson's whale. If all three encounters involved the same species, the observed differences may be due to sexual dimorphism or the normal variation between individuals or separated populations.

If the *BP Admiral* sighting was not of Clarke's whale, some cetologists suggested, it might involve the long-mysterious *Indopacetus pacificus*. A 1999 paper by cetologists led by Robert Pitman surveyed numerous sightings and photographs of a whale about this size with a prominent, falcate dorsal fin. Pitman suggested this whale, which bore no relationship to the smaller, differently marked Species A, was *I. pacificus*. At the time, Pitman called this species the tropical bottlenose whale. Once specimens of *I. pacificus* became available in 2002, Pitman was able to verify his tropical bottlenose was indeed this heretofore-mysterious species. However, Clarke's whale does not fit this picture, as Clarke had a clear view of his animal's head and saw there was no beak.

To recap this confusing mix of evidence, cetologists have solved one mystery by matching *I. pacificus* to the tropical bottlenose whale, but there is not a consensus yet on Species A, and Wilson's whale seems to be something else entirely. Clarke's whale might be the same species as Wilson's, or Clarke could have observed an oddly marked population of orcas (the essay "Wolves of the Sea," a little further along in this book, discusses the many interesting variations in both the genetic makeup and appearance of orcas). Then again, Clarke's whale could be still another unidentified species.

Karl Shuker has suggested there may be a new cetacean already in captivity. In a tank in Jakarta, Indonesia, are four *pesuts*—river dolphins from Eastern Borneo, captured in 1989. Until these specimens were captured, biologists reportedly dismissed the pesut as a myth. Now that it's in captivity, there's disagreement about what it is. Pesuts look like small members of the species *Orcaella brevirostris*, the Irrawaddy dolphin. However, the pesuts have some notable peculiarities. None of the captive specimens has any teeth, while the Irrawaddy dolphin may have more than seventy. Moreover, the pesuts have been reported to stun prey by forcing streams of water at it from their mouths. The prey is then sucked in, using a vacuum action, to be swallowed. All this makes Shuker wonder if the pesut is actually a new species. Apparently, genetic testing has not yet been carried out.

Next we come to the monodonts. Monodonts are members of a Northern Hemisphere whale family characterized by moderate size, specialized, shortened skulls, rounded flippers, and the absence of a dorsal fin. The only known members are the beautiful white beluga (*Delphinapterus leucas*) and the impressively "horned" narwhal (*Monodon monoceros*). Both species, it seems, have their mysteries.

In 1998, paleobiologist Darren Naish was watching an episode called "The Frozen Ocean" in the BBC documentary series *Kingdom of the Ice Bear*. On the film for about six seconds appeared a medium-sized Arctic whale with no dorsal fin. It was predominantly black with a distinctive white area on its side. This cetacean was swimming amid a school of belugas. Apparently, the filmmakers didn't notice the unusual whale: it seems to have been filmed incidentally and was not remarked on by the narrator.

Unfortunately, only the animal's back was filmed, so there's no way to know what the entire creature looked like. Still, enough was shown to demonstrate this was not an ordinary whale of a known species. The most famous black-and-white cetacean is the orca, but that species' dorsal fin is unmistakable, and no finless examples are known. The filmed cetacean differs in markings, morphology, or both from all other known dolphins and porpoises. What it looks like is a black-and-white monodont—something never before reported.

Both known species of monodonts are highly social creatures, and the two are often seen in association. If the black-and-white whale is an unknown, third, species, its proximity to members of a beluga pod would by no means be unexpected. Another possibility, perhaps the most likely, is that this animal is the first known partially melanistic beluga.

Naish has also researched another monodont-related mystery. As mentioned above, all known monodonts are Northern Hemisphere animals. However, a curious sighting occurred on an Antarctic expedition of 1892-93. Expedition leader W. G. B. Murdoch, in his book on the adventure, wrote that, on December 17th, 1892, several of his men caught sight of some narwhal-like whales. The passage describing this incident is very brief: "Just after killing the seal there was a shout amongst the men forward, 'A Uni! A Uni!'—the whalers' term for a Narwhale [sic]. Several men said they saw their horns."

Naish notes there are several peculiarities about this report. As 'horns' were mentioned, there must have been more than one of these whales. Since they were seen in Antarctic waters, they must have been cold-adapted animals. So, assuming the report is truthful, we are faced with a cold-adapted whale equipped with a horn. Could there be an austral narwhal? Presently, there is no evidence that narwhals ever lived in or anywhere near the Antarctic Ocean. Cryptozoologist Michel Raynal, who has uncovered two other isolated reports of horned cetaceans in the Southern Hemisphere, suggests that southern whales convergently evolved a narwhal-like form. No hard evidence for any such convergence yet exists. It's possible, but we have very little data to go on.

While unusually colored monodonts are so far unknown (or at least unconfirmed), abnormal coloration has been reported in many other cetacean species. Most startling are the all-white examples of whales which normally are dark-colored. At least two examples of albino sperm whales are on record, an adult caught in Japan and a calf presently living. (A white animal is not necessarily a true albino: several less-drastic genetic oddities can create white or predominantly white specimens.) A white humpback whale was first spotted in 1993 off Australia and has since been seen several times on its annual migration. White orcas named "Moby Doll" and "Chimo" and a white bottlenose dolphin named "Snowball" have been kept in captivity, and in 1994 another white bottlenose dolphin (*Tursiops truncatus*) was spotted in the Gulf of Mexico. In fact, anomalous white individuals have been reported in no fewer than twenty-one species of cetaceans.

Over seventy sightings of white orcas were reported from 1923 to 1959. In one case, an apparent albino female was seen with a calf which was nearly all white, but whose dorsal fin was fringed in black. In addition to the animals already mentioned, white cetaceans have included bowhead, right, sei, fin, and gray whales.

There are melanistic examples as well, including several species of dolphins. All these cases must be kept in mind when evaluating reports of possible unclassified cetaceans.

Whales are among the most fascinating animals on the planet Earth. It's both exciting and humbling to realize we still have so much to learn about them.

CRYPTIDS THAT NEVER WERE

In any examination of cryptozoology, it's important to report on the negative outcomes of investigations as well as the positive ones. Cryptids which turned out to have mundane explanations provide important cautions for cryptozoologists, some of whom can be too eager to accept new animals on the basis of insufficient evidence. These situations can also illustrate how the search for new animals should and should not be conducted.

Sometimes cryptozoologists have been taken in by hoaxes. The case of "Old Three-toes" was one example. In 1948, a clever jokester named Tony Signorini fashioned three-clawed iron feet and left tracks on the Florida beaches which fooled even veteran zoologist Ivan Sanderson. Signorini set out to make imitation dinosaur feet, but what he created were very close in appearance to penguin tracks, and Sanderson thought he was on the trail of the granddaddy of all seabirds. Signorini's friends helped out by instigating some fake sighting reports of the supposed monster, with rather ambiguous descriptions. The mystery was not definitively cleared up until the hoaxer confessed in 1988.

I wonder if Sanderson's acceptance of the Florida monster was partly due to his lack of familiarity with the area. I grew up in Florida, which has many types of beach sand. I remember one grayish kind which changed in consistency from quicksandlike to concrete-hard depending on the tide and the amount of sea water seeping through it. Signorini left his footprints at night, and Sanderson saw them during the day. If the tide had turned and the sand was much harder, this might have led to Sanderson's stated belief that the prints were too deep to be left by hoaxers.

Rivaling Three-toes for the "most famous hoax" title must be Francois de Loys' South American ape. As mentioned earlier, the 1917 photograph produced in support of de Loys' claim shows an animal that, while it has no visible tail, is far too much like a spider monkey (*Ateles belzebuth*) to be acceptable as a new species to most zoologists. The animal's widely separated nostrils, the tiny, curved thumb, and the triangular white patch on the forehead are only some of the telltale characteristics.

Loren Coleman and Michel Raynal, two cryptozoologists who have investigated this case, believe George Montandon, the French anthropologist who publicized de Loys' find, did so to support his own racist belief in a New World line of apes which led to the evolution of "lesser" peoples, like Native Americans. According to Raynal, there is some evidence de Loys was in on the hoax and that the animal was not even photographed on the expedition of 1917.

The case of the Silver Lake Serpent is unique—an apparent double hoax. A serpentine monster was reported several times in Silver Lake in New York state in 1855. In 1857, a fake monster was reportedly discovered in the attic of the lakeside Walker Hotel. The contraption of wire and canvas was floated by compressed air and pulled by underwater ropes.

In 1999, Joe Nickell, a skeptical investigator of anomalies of all types, discovered there was no proof the hoax monster had any more basis in fact than the "real" one. Nickell was unable to dig up the alleged confessions of hoaxers or any historical accounts of the finding of the "monster" remains. This convoluted (one is tempted to say serpentine) tale of misperception and misdirection is a caution to everyone about the acceptance of monster reports and of unverified explanations.

Wherever there is a mystery, there will be jokers. Loch Ness has attracted more than its share. Tracks have been made with a stuffed hippopotamus foot. Many photographs of the "monster" have been fakes. The most famous of Nessie's portraits, the 1934 Surgeon's Photograph, is allegedly a hoax as well, although the hoaxer's claim has been disputed. All this does not mean the creature itself is a hoax, but it does make proper investigation very difficult.

The ri from Papua New Guinea is another kind of case. Here a reported unknown animal was eventually unmasked as a known animal (the Indopacific dugong, *Dugong dugon*) displaying previously unknown behavior. The dugongs in this area flexed their backs more than other dugongs when diving and stayed submerged for longer periods than previously recorded.

When a sea creature called part human and part fish was described to anthropologist Roy Wagner by the Barok people in the early 1980s, experts put forth candidates including the dugong and assorted cetaceans. Marine biologist Kevin Britton suggested a beluga, an animal only known to live in the Arctic. Britton's theory was based on the existence of a single beluga skull which was allegedly collected off Australia. (Darren Naish has investigated this case and believes a museum mislabeled the specimen.)

What actually happened in the ri affair was that the most proper type of investigation was pursued. The case began with local reports of an animal which interested scientists could not immediately identify. Some cryptozoologists, on expeditions led by Wagner and later by Thomas Williams, went out and looked at the animal for themselves. They also studied the local folklore and languages, trying to identify the origin of the terms ri and *ilkai* (the name given the same animal by another tribe, the Susurunga). It is worth noting this was a case in which the knowledge of the Native peoples was incorrect. According to both of the tribes involved, the ri and the dugong were different animals. The result: while no new species was found, some knowledge was added to science.

The moral of this stories is that proper scientific investigation is rarely a waste of time, whether it results in a spectacular new animal or not. Speculation on the basis of insufficient or unexamined evidence, on the other hand, usually is fruitless. A science as controversial as cryptozoology requires a careful balance between open-mindedness and skepticism, with the most promising cases being subject to a cautious and thorough examination.

AUSTRALIA'S SHADOW PREDATORS

As everyone knows, Australia houses perhaps the most unique fauna on Earth. In recent years, some new zoological questions have emerged on this continent. Do reports of a "puma" or "cougar" arise from mistakes and hoaxes, or is Australia host to naturalized American felines? Did the introduced dingo truly drive the marsupial thylacine to extinction? And is there, or was there, a catlike predator in Australia's forests?

In 1994, the *North Central News* in St. Arnaud, Victoria, ran a grisly photograph of an eviscerated sheep which a farmer believed had been killed by a "puma." Sightings and footprint casts added to the evidence for some kind of large predator. In the same region, in 1985, a Forest Commission officer cast tracks of a cat which leaped a creek over fifteen feet wide. No known Australian predator is capable of such a feat, which brings to mind the legendary long-jumping abilities of the American puma. In November 1999, Russell Robinson of North Drummond found his horse, Jack, with severe wounds. These had been inflicted by a large animal which was clearly perched on top of the horse, an assault impossible to ascribe to any creature belonging in Australia.

In their book *Out of the Shadows: Mystery Animals of Australia*, Tony Healy and Paul Cropper reported a story of a female puma with four cubs brought over as mascots by an

Unidentified Australian predator photographed by Ms. Rilla Martin.

American fighter unit and dumped in the wilderness in 1943. Another story suggests American gold miners brought pumas with them in the 1850s. Zoologist Malcolm Smith, author of the other good book on Australian mystery animals, *Bunyips and Bigfoots*, disputed the "mascot" stories, but maintained it is "undeniable fact" that something alien to Australia's known fauna is killing stock on a large scale.

The authorities are beginning to agree. An inquiry by the state government in New South Wales reported in November 2003 that it is "more likely than not" that a breeding population of some form of "exotic large cat" existed in the state. The inquiry, begun after a teenager from Kenthurt said he'd been attacked by a "panther" and had deep slash wounds to prove it, was summarized by Dr. Johannes Bauer, a cat expert retained as consultant, who said, "Difficult as it seems to accept, the most likely explanation…is the presence of a large feline predator."

The whole situation is strikingly similar to that involving the supposedly extinct Eastern puma of the United States. There are plenty of believers and plenty of evidence, but the definitive proof —the animal itself, living or dead—has yet to be obtained. Also, many sightings, including one by a former British zookeeper and one by a police officer who saw a big cat jump right over the hood of his car, describe a "black puma"—a subject as controversial in Australia as it is in the United States.

In the summer of 2000, video footage shot in the Grampian Mountains in Victoria thrust the mystery into national prominence. As described in one news account, the tape showed "a sleek, muscular dark-skinned cat similar to an American puma loping through long grass." The video was not quite definitive: experts disagreed about the apparent size of the animal, with some saying it could be a very large feral domestic cat. Rob Wallis (the zookeeper mentioned above) once cast a large cat print in the same area. The print was identified at the Melbourne Zoo as most likely being a puma's.

In Queensland, a debate continues about whether Australia has a marsupial "tiger," known to Aborigines as the *yarri*. This question is so intriguing it is dealt with separately in the next essay.

Also figuring into this puzzle is the more doglike thylacine, discussed in Section II of this book. The thylacine presumably died out in Australia, perhaps as long as 3,000 years ago, but some evidence indicates it may survive. In Western Australia, east of Derby, the humerus bone of a thylacine was found in 1970. One zoologist, Dr. Michael Archer, believed this to be less than eighty years old. Further south, a dead thylacine was discovered in a cave in 1966. This carcass, still covered with fur, was carbon-dated at 4,600 years. This estimate, while accepted by most experts, is disputed by naturalist Athol Douglas. Douglas believes the animal had been dead less than a year when it was found. He visited the cave and found a dingo carcass, less than twenty years old, which had decayed much more than the supposedly far older thylacine. It should be mentioned there were other species of thylacinids—perhaps ten or more—besides the well-known Thylacinus cynocephalus. *Thylacinus potens*, for example, was built like the modern thylacine but was heavier, weighing about ninety pounds.

The most recent sightings which specifically describe a thylacine or similar animal come from Queensland, where the animal is called "The Beast of Buderim" or "The Tassie Tiger," depending on location. But it was far to the southwest, in Victoria, where the most intriguing piece of evidence appeared. In 1964, Ms. Rilla Martin took a single photograph of a striped animal. In the black and white photo, the animal's head is partially obscured by brush. The beast looks something like a thylacine, but there appear to be significant differences, especially in the shape of the hindquarters and tail. Is this some form of mainland thylacine relative, a hoax, or perhaps something altogether different? No one is certain.

Witness descriptions disagree in many of these cases, as do the footprints. What is clear is the existence of enough evidence to justify continued inquiry, which just may lead to some very interesting discoveries—or even sensational ones.

WOLVES OF THE SEA

There are few mammals more interesting to watch or to study than dolphins. There are many taxonomic questions about these mammals, but the most interesting concern the largest member of the family Delphinidae, the killer whale or orca. (The traditional, if prejudicial, term "killer whale" seemed on its way to extinction in the late 20th century, but is now coming back into use.)

Until the last few decades, when humans got to know this animal at close range, our species saw nothing to relate the malevolent-looking orca to its small and playful kin. We now know that orcas are indeed smart, curious animals with no antipathy toward humans, but they are certainly not creatures to be trifled with. A large male killer may be thirty feet long, with a dorsal fin six feet high. With their power, intelligence, and sophisticated pack hunting techniques, killer whales can lay legitimate claim to the title of "top predator" in the ocean ecosystem. Even the largest great white shark or bull sperm whale will vacate the area when a pod of orcas swims into view.

Traditionally, all orcas were believed to be a single species, *Orcinus orca*, with a worldwide distribution. Today, many experts think there may be two species—or three, or four, or even more.

One such expert is Dr. Robin Baird, a cetologist in the biology department of Nova Scotia's Dalhousie University. After many years of studying the killer whales of the northeastern Pacific, he suspects they include species yet to be described.

There are two distinct populations of killer whales in this area. These are known as the "residents" and "transients." The observable physical differences are small. The transients have more pointed, centrally located dorsal fins and larger, more uniform gray "saddle" patches behind their dorsals. Still, many animals which are distinct forms—even separate species—likewise have only minor differences in appearance.

To Baird, the available evidence in every area (behavioral, morphological, ecological, and genetic) supports his belief the two populations are reproductively isolated. Even the animals' diets are different. While the same prey species are available to both populations, the residents eat primarily fish, while transients are specialists who hunt marine mammals, including large whales.

Baird suggests these two forms are, at least, incipient species—

that is, populations on a divergent path which will eventually lead to distinct species. They may be species already. No one, however, has formally taking the step of designating a type specimen and naming a new species. Among other problems, Baird notes, no one knows which species would be the "new" one.

Just to make the orca question even more intriguing, there now appears to be a third population, the "offshore" orcas. The offshore animals, also first identified in the northern Pacific, live in large pods, up to sixty animals, and are primarily fish-eaters. They appear to be slightly smaller than the other types and have dorsal fins with rounded tips. They overlap in their range with both residents and transients but, so far as is known, keep to themselves.

The situation thousands of miles away at the bottom of the world is surprisingly similar. Two separate teams of Russian scientists in the 1980s described two new species of orca from the Antarctic. These two species, *O. nanus* and *O. glacialis*, are reportedly similar to each other (both are relatively small, with yellowish undersides) and may in fact be one and the same, although described from different locations.

At the time these descriptions were published, most cetologists considered them unverified. Dale Rice, author of the authoritative 1998 taxonomic guide *Marine Mammals of the World*, wrote that he did not entirely dismiss *O. glacialis*. He noted the skulls and teeth referred to by the describers, A.A. Berzin and V.L. Vladimirov, did indeed differ from those of common orcas. Reportedly, the smaller cetaceans cruise the same waters as members of *O. orca* but do not associate.

The claims about *O. nanus* and *O. glacialis* are hard to check because the describers of the former failed to identify a type specimen (making the name invalid to begin with) and the type specimen for the latter has apparently been discarded by the museum in Vladivostok which held this important material.

Despite this confusion, there may well be something to the idea of different species. Dr. Robert Pitman's experience with Antarctic orcas led him to think there are definitely three forms, and these may well represent species. His observations initially indicated two populations, a small fish-eating type that roamed the pack ice at the southern edge of the Ross Sea, and a large mammal-eating kind which preferred open water. As he examined the small whales further, he found they could be divided into two groups by appearance and behavior. Among other peculiarities, pack ice whales living near the Antarctic Peninsula preferred seals as prey. These pods displayed a number of clever, cooperative methods for tipping, grabbing, or washing seals off ice floes.

Pitman attacked the problem, not just by making notes and taking photographs, but by chasing whales through the pack ice in an open boat and using crossbow-fired darts to get small samples of skin and blubber for DNA analysis. (Fortunately, the formidable mammals did not object to this form of inquiry.)

In late 2003, Pitman published a paper putting forth the claim of three distinct types, which he suspects but cannot yet prove are biological species. His Type A is the "basic" *Orcinus orca*, but Types B and C are smaller, and are distinguished from Type A and each other by behavior, diet, range, and appearance, most easily noticed in the white eye patch (a different shape for each type). Pitman has yet to see any intermediate whales that might indicate the three populations cross-breed. Types B and C are both marked unusually for orcas, with a distinctive gray "dorsal cape" beginning as thin lines flowing from behind the head. The term "cape" is apt: the gray areas join behind the fin to make a very prominent marking, with a light band like a collar overlaid by the darker cape. The effect is to make the animal look as if, indeed, it has dressed for a night at the opera.

Analysis of Pitman's DNA samples and those from other populations are not complete as of this writing. No one knows what further study will reveal. The genetic fault lines running through the global population we now know as *Orcinus orca* are deep yet subtle. It might eventually be concluded that killer whales constitute nothing more than an unusually variable species. It seems more likely, though, that an animal which has been so successful and pushed into so many areas of the oceans has had plenty of opportunity to diverge into species. Robin Baird suggests orcas may even be the "Darwin's finches of the marine mammal world." Sorting out the facts on these magnificently lethal predators will keep cetologists busy for a long time to come.

THE YARRI

There is one peculiarly empty niche in the zoological web spanning Australia's deserts, forests, plains, and mountains. There are large numbers of sizable herbivores (the kangaroos), but no corresponding top predators. The dingo was introduced by early humans, so it doesn't count. The thylacine is believed to have died out about 3,000 years ago, although, as we have seen, there is some dispute about this. Is there, or was there, anything else filling the occupation of kangaroo predator?

Some aborigines in Queensland always thought there was. They called it the *yarri*, and described it as a puma-sized striped carnivore totally unlike the dingo or the thylacine. From the 1860s on, white settlers and explorers also saw it on occasion. Invariably, they described it as resembling a "cat" or "tiger." The first notice in a scientific journal about this animal was published in the *Proceedings of the Zoological Society* of London in 1871.

It is all too common in cryptozoology for a reports to come in for decades without any credible-sounding claim of a carcass being found. In the yarri's case, there were several reported killings between 1900 and 1932, although none of the carcasses was ever provided to a museum or other institution. Naturalist George Sharp, for example, wrote that he once examined a skin almost five feet long from a cat killed near Atherton.

An authoritative 1926 book, *Wild Animals of Australasia*, matter-of-factly described the "North Queensland striped marsupial cat" as fawn or gray with black stripes on its flanks. The authors, A. S. le Souf and Harry Burrell, noted that most reports also indicated the animal had "sharp, pricked ears" and was about five feet in total length.

The fossil record of Australian carnivores offers some clues to the yarri's possible identity. At one time, there were many large predators on the continent. The most famous was *Thylacoleo carnifex*, a marsupial "lion" that may have weighed nearly as much as its placental African namesake. Some reconstructions of this animal (of which we have a complete skeleton) are very catlike: others suggest a sort of giant, murderous possum. Whatever its exact appearance, Thylacoleo was a fearsome creature, built for power rather than for speed. It had heavily muscled forelimbs like a jaguar, oversized claws on semi-opposable thumbs, and teeth "like bolt-cutters," as paleontologist Stephen Wroe put it. There were several smaller relatives of this animal, some in the genus Thylacoleo and some assigned to Wakaleo, which may have been more leopardlike and arboreal. At least one species survived as long as 50,000 years ago, and there are two aboriginal paintings which appear to depict striped catlike beasts.

In short, there are plenty of candidates for the yarri's identity. Everyone's sentimental favorite is *T. carnifex*, but this animal may have been too large to fit the descriptions of the yarri. In one account, a yarri was described as "about the size of a dingo." Sharp's cat would have stood only about eighteen inches high in life.

Most experts who have written on the subject concur that witnesses have not mistaken surviving mainland thylacines—assuming these exist—for yarris. There is one drawing of a yarri track, which looks rather catlike and clearly was not made by a thylacine.

In addition, several witnesses have remarked on how the yarri's front teeth protrude in a fashion observers have compared to tusks or even to the fangs of extinct sabre-tooth felines. This is totally unlike a thylacine, but, as Karl Shuker points out, it is a trait shown in Thylacoleo skulls.

Writing in 1952, Dr. Maurice Burton, Britain's most eminent zoologist, expressed his opinion the yarri might well be a real animal. Noting the large virgin forest tracts in Queensland, Burton wrote, "…it is not surprising if no zoologist has seen a 'tiger-cat' and no camera has recorded it. It is merely another instance of random observation giving the only chance of success."

Reports have declined in frequency since Burton wrote, but there have been some solid ones. In 1987, for instance, a hunter claimed he had wounded a dingo (the wild dogs are often killed to protect livestock) but lost his quarry when a striped cat pounced on the dog and ripped it apart. A dentist near Buderim in 1994 described a very clear sighting of a large tigerlike animal which froze in his car's headlights and stared at him for about half a minute.

In 1992, Colin Groves, an authority who is always cautious about cryptozoological claims, expressed his doubts that feral domestic cats and other mundane explanations could entirely account for the yarri reports. He wrote, "The Queensland Tiger has always seemed to me to need further scrutiny: those reported fangs are suggestive of a specialized carnivorous marsupial."

There have been no reports of carcasses for some seventy years, however. It is possible the yarri—whatever it may be—is a declining or recently extinct species.

We leave this mystery in a most unsatisfying state. The clues offered by ecology and paleontology make the yarri seem quite believable. Only the lack of hard evidence stands in the way of its confirmation. At the least, the yarri seems one of the most promising candidates for future cryptozoological expeditions.

A WHALE OF A HYBRID

Unusual cetaceans are most often thought to be anomalous individuals of a known species or members of an unknown species. In some cases, there is a third alternative: the animal might be a hybrid. There are a surprising number of cases in which different species of cetaceans have interbred, with some striking results.

For example, the skull from a whale killed in Greenland in 1986 or 1987 appears to be evidence of a hybrid between the two known monodonts, the beluga and the narwhal. The skull was spotted in 1990 by Mads P. Heide-Jorgensen of the Greenland Fisheries Research Institute. It was sitting on the roof of a tool shed in the settlement of Kitsissuarsuit.

The skull belonged to a hunter named Jens Larsen. Larsen had who killed three identical whales, one of which produced the mystery skull. He recalled that the animals seemed very strange to him. They were a uniform gray color, showing neither the distinctive white of a beluga nor the mottled back of a narwhal. Their tails looked like a narwhal's, with their distinctive fan-shaped flukes and convex trailing edges. Their broad pectoral flippers, though, resembled a beluga's. While these cetaceans had no horns, analysis of the skull indicated two teeth showed growth patterns resembling the spirals of a narwhal tusk. These teeth may have protruded outside the mouth.

Hybridization is also known to have occurred between the two largest animals on Earth, the blue whale (*Balaenoptera musculus*) and the fin whale (*B. physalus*). In a 1998 article in *Marine Mammal Science*, Martine Berube and Alex Aguilar reported finding five such examples documented in scientific literature. The authors devoted most of the article to a hybrid caught off Spain in 1984. This animal showed features intermediate between the two parents. The whale was four years old, and, at sixty-three feet in length, "anomalously large" for its age. Another instance, described in the *Journal of Heredity* in 1991, concerned a similar hybrid caught off Iceland in 1986. In the 1986 case, the whale turned out to be a pregnant female. Analysis of the fetus indicated the father was a blue whale. This was the first case in which such a hybrid was proven to be fertile.

These hybrids tend to be dark gray and are usually mistaken for fin whales. Curiously, an author named A. H. Cocks, writing in 1887, stated there were three kinds of giant baleen whales caught off the Norwegian coast. One type appeared intermediate between the blue and fin whales, so much so that Cocks called it the "bastard" and suggested all such animals were hybrids. If this is an accurate deduction, hybridization between these two species has been going on for a long time.

Such a hybrid has figured in the controversy over the continuation of whaling for "scientific research" in Japan. While meat from the minke whales (*Balaenoptera acutorostrata*) taken in this program is sold legally in Japanese markets, the government has always denied that meat from any protected species is sold or imported. This was disproved when a sample purchased in a Japanese market in 1993 was traced to a specific whale killed off Iceland in 1989. This was possible because the whale involved was so rare—a blue/fin hybrid. (One wonders at the competence of Japanese whalers who, in the killing of hundreds of whales a year, have apparently not developed the expertise to tell a relatively diminutive minke from much larger species.)

Such blue/fin hybrids are still reported off Iceland. A recent sighting near that island nation is even mentioned in promotions for a whale-watching tour company.

More recently, a new kind of hybrid was reported. A calf spotted near Tahiti in 2000 with its mother, a humpback whale, looked like a hybrid between a humpback and a blue. The calf was abnormally large, yet with pectoral fins that were shorter than normal for its species (the humpback is the only member of its genus, Megaptera, which is named for the whale's oversized fins). The calf also displayed the coloration of a blue whale. The parents are not just from different species but different genera. Intergeneric hybrids are freakishly rare and always unexpected. Mammologist Michael Poole speculated that, with blue whales still rare and thinly spread out over the oceans, loneliness may have driven a male blue to seek an unusual mate.

In a Japanese aquarium, a male Risso's dolphin (*Grampus griseus*) produced calves with three different bottlenose dolphins (*Tursiops truncatus*). In 1979, the Whales Research Institute in Tokyo used a photograph of one of these intergeneric calves on a Christmas card. Richard Ellis, who is well-known as both a writer and an artist specializing in cetaceans, wrote, "...when I opened the envelope and saw a shiny gray, short-beaked cetacean, I was struck dumb: I had been studying pictures of these animals for years, and before me was an animal I couldn't even begin to identify." In 1981, another aquarium in Japan reported a male false killer whale (*Pseudorca crassidens*) and a female bottlenose had produced a calf.

While the blue whale-humpback whale mentioned above is the most spectacular example of an intergeneric cross reported in the wild, there have been a few other incidents. Robin Baird wrote in 1998 that a fetus recovered from the corpse of a Dall's porpoise (*Phocoenoides dalli*) proved to have an unusual father: a harbor porpoise (*Phocoena phocoena*). Baird found this particularly intriguing because there are several other reports of unusually pigmented cetaceans with the general size and form of Dall's porpoises. Although Dall's porpoises are notably variable in their pigmentation, Baird suggests some of these cases are due to ongoing hybridization with harbor porpoises. Another intergeneric hybrid, this one between the long-beaked dolphin (*Delphinus capensis*) and the dusky dolphin (*Lagenorhynchus obscurus*) was nabbed off Peru.

In 2001, an apparent hybrid between a dusky dolphin and a southern right whale dolphin, *Lissodeplhis peronii*, was photographed among a school of duskies. This very unusual-looking animal was about seven feet long, larger than normal for a dusky. It sported a solid black upper body and was completely white underneath, lacking the intermediate shades normally present on a dusky. On the other hand (or flipper), it had black pectoral fins, whereas the right whale dolphin's are white, and it had a small triangular dorsal fin. Right whale dolphins have no dorsal fin at all.

Finally, three odd-looking dolphins which washed up on an Irish beach in 1933 were identified by one expert as hybrids between the bottlenosed dolphin and Risso's dolphin. While the match between these two species was proven viable by the incident from captivity described above, not all cetologists accept the hybrid interpretation in this case.

The popular bottlenose seems to be the main instigator of hybridization in captivity. One paper recorded twenty-one incidents in which a Tursiops mated successfully with another species. It's true there are more bottlenose dolphins in captivity than any other cetacean, but that's still quite a surprising record. To indulge in a completely illogical, yet human, thought, it's enough to make people wonder what's behind this animal's famous "smile."

THE SECRET OF LAKE ILIAMNA

Alaska is a land of countless lakes, many of them impressively large. Lake Iliamna, however, is like no other. It might be the grandest physical feature in the United States that most American citizens are completely unaware of.

Fully eighty miles long and with a surface area over a thousand square miles, Iliamna is approximately the size of the state of Connecticut. Iliamna is the seventh-largest body of fresh water in the U.S., with a mean depth of 144 feet and a maximum depth greater than 900 feet. The lake is connected to Bristol Bay, sixty miles southwest, by the Kvichak River, through which such marine mammals as harbor seals and belugas can travel. Iliamna, in fact, supports a resident population of harbor seals, along with a successful sport-fishing industry.

The most intriguing thing about Lake Iliamna, however, is the possibility it houses unknown animals of enormous size. The lake's mysterious denizens are nothing like the classic long-necked "lake monsters" alleged to dwell in other bodies of water. Instead, the animals reported from Iliamna look like gigantic fish.

Reports of something odd in Iliamna began with the indigenous tribes, although no one knows how long ago such stories first took root. The Natives did not hunt the lake's creatures, and believed them to be dangerous to men fishing in small boats. While some early white settlers and visitors reportedly saw the things, too, stories about Iliamna did not gain wide circulation until the 1940s, when pilots began spotting strange creatures from the air. The flyers' descriptions generally matched the tales of the Alaska Natives. The lake's mystery inhabitants were most often described as long, relatively slender animals, like fish or whales, up to thirty feet in length.

In 1988, "Babe" Alsworth recounted his 1942 sighting in an interview with cryptozoologist Loren Coleman. Alsworth, a bush pilot and fishing guide, saw a school of animals well over ten feet long in a shallow part of the lake. Alsworth described them as having fishlike tails and elongated bodies and described the color as "dull aluminum." Larry Rost, a survey pilot for the U.S. government, saw a lone creature of the same type as he crossed the lake at low altitude in 1945. Rost thought the animal was over twenty feet long.

There have been at least three attempts to find or catch Iliamna's mystery inhabitants. In the 1950s, sportsman Gil Paust and three companions (one a fisherman named Bill Hammersly,

Aerial sighting of mysterious fish-like giants of Lake Iliamna.

who had been in Babe Alsworth's plane in 1942 and shared in that sighting), tried to fish for the creatures. According to Paust, something grabbed the moose meat used as bait and snapped the steel cable it was hooked to. In 1959, oilman and cryptozoology enthusiast Tom Slick hired Alsworth to conduct an aerial search of the lake, but nothing was sighted. An expedition in 1966 also apparently met with no success, as no results were announced.

In 1979, the *Anchorage Daily News* offered $100,000 for tangible evidence of the Iliamna creatures. The reward attracted both serious and eccentric researchers (one man reportedly played classical music to lure the animals up). Apparently, there has never been a well-financed expedition with sophisticated sonar and photographic gear.

According to a 1988 article in *Alaska* magazine, a noteworthy (but unnamed) witness was a state wildlife biologist. In 1963, this official was flying over the lake alone when he spotted a creature which appeared to be twenty-five to thirty feet long. In the ten minutes it was under observation, the animal never came up for air. Other flying witnesses mentioned in media accounts include a geologist who flew over the lake with two companions in 1960, reportedly spotting four ten-foot fish, and Tim LaPorte, who reported a sighting in 1977.

In LaPorte's case, the veteran pilot and air-service owner was near Pedro Bay, at the northeast end of the lake. He was flying just a few hundred feet above a flat calm surface. LaPorte and his two passengers, one a visiting Michigan fish and game official, saw an animal lying still, its back just breaking the surface. As the plane came closer, the creature made a "big arching splash" and dove straight down. LaPorte still remembers watching a large vertical tail moving in slow side-to-side sweeps as the animal sounded. Comparing the object to a familiar type of eighteen-foot boat seen from the same altitude, LaPorte and his companions estimated the thing was twelve to fourteen feet long. LaPorte described the creature as either dark gray or dark brown. LaPorte had been a passenger in a different aircraft in 1968 when the other two individuals in the plane had a very similar sighting. (In that incident, LaPorte, who was taking flight instruction and sitting in the left seat, could not see the animal from his side.)

Modern sightings have occurred mostly near the villages of Iliamna and Pedro Bay. It was off the latter town in 1988 that several witnesses, three in a boat and others on shore, reported one of the creatures. In this case, it was described as black. One witness thought she could see a fin on the back, with a white stripe along it.

It is important to note that Lake Iliamna today remains isolated, its shores largely unpopulated. The largest village, Kakhonak, counts only 200 permanent residents. There is no highway to provide easy access to (or from) the outside world. Sport fishermen and other summer visitors come by boat or fly in to the area's few airstrips. If there are unusual creatures in the lake, it's hardly surprising that a long time can pass between good sightings.

A common theory about the Lake Iliamna creatures (sometimes called "Illies") is that they are gigantic sturgeon. These could be either an outsized population of a known type or an unknown species. Sturgeon—large, prehistoric-looking fish, with armorlike scutes covering their backs and a heritage dating back before the dinosaurs—match most descriptions from Iliamna fairly well. Sometimes the match is precise. Louise Wassillie, who watched a creature from her fishing boat in 1989, said, "It's only a fish. It was about twenty feet long and had a long snout. Probably a sturgeon." An earlier witness named Eddie Behan told writer Kim Fahey of seeing a twenty-foot spindle-shaped animal with a fish's tail and rows of lumps on its back—a good description of what a sturgeon that size would look like.

Biologist Pat Poe of the Fisheries Research Institute (FRI) at the University of Washington, who has studied the salmon populations in Iliamna and neighboring Lake Clark, once commented, "I'm sure there's a big fish. I think the lakes have a lot of interesting secrets. We don't know much about other resident fish in the lake." Warner Lew, currently the senior biologist with the FRI's Alaska Salmon Program, agrees the lake seems a suitable habitat for large sturgeon. Lew reports several witnesses have told him of sighting giant fish, but he has yet to see any fish larger than a four-foot Northern pike in his twenty-four years of research visits to the lake.

There are nine species of sturgeon in North America. The white sturgeon (*Acipenser transmontanus*) is the largest of these, and is the continent's largest fresh-water fish. The record claim for a white sturgeon was made for a fish caught in Canada's Fraser River in 1912. The sturgeon was twenty feet in length and weighed 1,800 pounds. A fish of 1,500 pounds was reported caught in 1928 in the Snake River in the United States. An eleven-foot specimen weighing 900 pounds was found dead on the shore of Seattle's Lake Washington in 1987.

Sturgeon expert Don Larson, curator of the Sturgeon Page Website, reports sturgeon over ten feet long are often caught in the Fraser and Columbia Rivers. Larson comments, "Most biologists I've talked to say that white sturgeon over twenty feet and 1,800 pounds is highly probable."

White sturgeon are not known from Iliamna, but have been found in other Alaskan lakes and in coastal waters as far north as Cook Inlet. There is a single record of a catch in Bristol Bay, which puts a migration to Iliamna within the bounds of possibility. It's also possible that white sturgeon became trapped in the lake thousands of years ago, when the last glaciers receded, and have developed in isolation. Jason Dye, a biologist for the Alaska Department of Fish and Game's Bristol Bay office, recently said, "There's never been any documentation that anyone's caught one in the lake, or seen one, as far as I know. But that doesn't mean they're not in there."

Sturgeon are bottom-feeders and would rarely be seen near the surface, which fits the Iliamna phenomenon. The appearance of white sturgeon—gray to brown in color, with huge heads and long cylindrical bodies—appears to match most Iliamna reports. (No one is certain how the species got the name "white sturgeon," although some genuinely white specimens have been reported from salt water.)

It may be a distinct sturgeon population has developed, distinguished from the known white sturgeon mainly by unusual size. Whether this hypothetical type is different enough to be a new species is unknown. There is plenty of food in Iliamna, where up to twenty million sockeye salmon return to the lake from the sea every year. (There is serious concern among conservationists that this number is declining, and that the salmon run is being too heavily depleted by legal and illegal fishing as the fish migrate via the Kvichak.) There is also plenty of room. Iliamna has fifteen times the volume of Loch Ness. At the same time, it must be admitted there is no physical or film evidence for unknown creatures of any kind.

A landlocked population of fish becoming larger than their relatives which are anadromous (dividing their lives between fresh and salt water) would be unusual. In most cases where a species has become split between freshwater and anadromous populations, as with salmon, the freshwater dwellers run smaller. However, this rule may not be valid for Lake Iliamna, with its huge size and bountiful food supply.

Sturgeon? Monster? Folklore? Or something completely different? Whatever is going on in Lake Iliamna, it makes for one of the most unusual and intriguing mysteries in the animal world. If any of the lake monster cases turns out to involve a real creature of prodigious size, it is Iliamna, not the more famous lakes in Canada and Scotland, where I would place my bet.

A KINGDOM OF MYSTERIES

We have only touched on the riddles of the animal kingdom. As we saw in Section I, there are likely millions of species waiting to be discovered. By sheer chance alone, some of those will undoubtedly prove to be large, bizarre, scientifically interesting, or maybe all three.

There are plenty of hints about other creatures which may be lurking in the forests and mountains of the world. From Africa's alleged flying reptilelike animal, the *kongamato*, to the giant anaconda-like snake reported for centuries from the Amazon basin, there is no shortage of wild wildlife stories, scraps of evidence, and unanswered riddles.

Take Beebe's manta, for instance. Pioneer ocean explorer William Beebe, inventor of the bathyscaphe, reported several fish no one has seen since. Most of these encounters occurred under water, but in April 1923, a strange-looking manta ray, measuring an estimated ten feet across the wing-like pectoral fins, actually ran into a ship he was on board. As the stunned fish revived and swam off, Beebe saw it was dark colored, like known mantas, but with significant differences. The fish had white wingtips and striking white bands running along what, on an airplane, would be

The strange-looking manta reported by William Beebe and others.

called the "wing roots," extending halfway down the sides of the body. Beebe felt this was a new species: in addition to the white markings, the trailing edges of the wings were nearly straight, not concave as is usual. The cephalic "horns" were straighter than normal, and the tail unusually short.

A similar manta, with some difference in the markings (the white bands flared out across the wings instead of running down the body as far as on Beebe's specimen), was filmed off Baja California in 1989. In 2003, the television game show *Survivor* debuted its fall season in the Pearl Islands west of Panama. On the opening credits for the first show was a very clear video shot of a strange manta, with white markings similar to those on Beebe's fish but running the full length of the body.

Scientists are still trying to make sense of a sound that turned up recently when civilians were given access to the U.S. Navy's formerly classified network of submarine-sensing hydrophones. Amid the chatter of dolphins, the clicking of shrimp claws, and the deep songs of whales was a noise given the highly descriptive name "Bloop." Marine biologist Phil Lobel, one of the scientists who have studied Bloop, agrees it sounds organic in origin but is too powerful to be generated by any known creature—even a whale. Giant squid have no apparatus to generate sound. So what is the source of Bloop?

On land, mammals in particular offer a number of mysteries—not quite as strange as Bloop, but still intriguing.

For example, there are several alleged African mystery cats, but the best reports come from the mountainous areas of Kenya. The beast involved is a kind of small lion covered with prominent brown spots. The male's mane is sparse and insignificant. To the local Kikuyu people, the spotted lion is the *marozi*, an animal distinct from the ordinary lion, *simba*.

The strong local tradition concerning the marozi is backed up by reports of European officials like game warden R. E. Dent, who saw four small spotted lions cross the path ahead of him in 1931. A short time later, a farmer named Michael Trent shot two such animals, a male and a female.

Naturalist Kenneth Gandar Dower obtained the skin of Trent's male lion and one skull. This evidence was examined by a prominent mammologist, Professor Reginald I. Pocock of the British Natural History Museum.

Pocock reported the dead cat was smaller than normal for East African lions, being about the size of a normal adolescent lion a year or so short of being full-grown. The mane was very sparse. The skull definitely belonged to a juvenile, either a female or an

Mokele-mbembe, Africa's reported "dinosaur."

unusually small male. It was the large, jaguarlike spots on the hide which made this "a remarkable specimen." Most lions have faint spots when they're born, but these fade quickly and never form the distinct pattern shown by Trent's skins. Pocock had previously examined two other skulls (apparently now lost), which he described as a male's and a female's, apparently belonging to marozi. These were from adult animals but were smaller than the smallest "ordinary" lion skulls.

The marozi, if it exists, is certainly rare. Indeed, no good reports have been made from Kenya since the 1950s, and the whole business might be dismissed as folklore if not for the evidence still available for examination.

The skin of the female specimen, along with a single skull, remains housed in The Natural History Museum in Britain. The curator, Daphne Hills, reports the skull on file belongs to a young animal of unknown sex. Ms. Hills believes too much has been made of the skin. She agrees the spots are unusual, but suggests that this merely represents an aberration. Other zoologists, such as Dr. Karl Shuker, continue to disagree, with Shuker pointing to the long-standing local traditions and other sightings of adult spotted lions.

The mystery cat called mngwa *or* nunda, *feared in East Africa.*

Colonel William Hichens, a British colonial official in Tanzania in the 1930s, wrote about his fruitless pursuit of a mystery cat the indigenous inhabitants called *mngwa* or *nunda*. This lion-sized, brindled gray feline was blamed for killing several people, including two policemen from whose bodies Hichens collected unidentifiable gray hairs. He sent these for expert analysis, but received only a rather unhelpful report concluding they were "probably cat."

Hichens admitted tales of strange animals "lose nothing in the telling," but argued, "a stretcher-load of clawed, mauled, and mangled man dumped at one's tent-door is no myth at all." After about a month, the killings in this episode simply stopped. There have been other reports like this from East Africa, but no one has caught any animal, known or unknown, identified as the culprit.

Southeast of Africa proper, the island of Madagascar has its own mammalian puzzle. Madagascar's top carnivore is the scrappy wildcat-sized fossa, *Cryptoprocta ferox*. Intriguingly, there is a strong local tradition concerning a larger carnivore that may be the fossa's prehistoric big brother, the panther-sized *Cryptoprocta spelia*. The world's leading fossa researcher, ecologist Luke

Dollar, found the reports convincing to the point he mounted a 1999 expedition into the so-called Impenetrable Forest of northeastern Madagascar in search of the animal. He did not find any evidence, but he still believes the creature's existence is possible. Dollar notes this area is so unexplored there are no maps of much of the forest's interior.

There is no easy explanation for the enormous bat zoologist/writer Ivan Sanderson and naturalist Gerald Russell reported from the Assumbo Mountains of western Africa in 1932. Sanderson insisted there could be no mistake, because the animal startled him half to death by swooping directly at him. The witnesses thought the bat's wingspan would measure at least twelve feet.

Nor is there any new information concerning the "ruffed cat," an animal whose skins Sanderson reported finding twice in marketplaces in Mexico's rural state of Nayarit. The larger of the two skins was about six feet long, not counting the tail, and came from an animal with a short muzzle and long legs. The flanks were striped light and dark brown, and long hairs formed a prominent ruff around the neck. The two skins Sanderson purchased were lost in a flood, and no specimen has since come into the hands of any scientist.

It may be relevant to note there are at least two photographs of adult or subadult pumas which have retained their juvenile spotting pattern. Dr. Marcella Kelly captured one such photograph in a night camera trap in Belize in August 2002. While a spotted puma is an interesting oddity to note on its own merits, its relationship to Sanderson's cat is unclear, as it wouldn't explain the ruff.

Another unusual mammal, which definitely exists but whose ancestry was at first a puzzle, was located in a valley in Nepal's Bardia National Park. In 1992, a British expedition obtained photographs of two extremely large and very strange-looking elephants. They appeared to be Asian elephants, but were over eleven feet tall, which would be a record for that species. They had weird-looking heads with high domes on top and prominent nasal bridges above their trunks. These latter features are associated with an extinct prehistoric form called *Elephas hysudricus*, and a television documentary on the creatures was titled *The Lost Mammoths*.

A follow-up expedition shed more light on the mystery. The discoverer of these elephants, explorer John Blashford-Snell, led another group into the remote park region and collected more film along with dung samples. Examination of DNA found in the animals' dung indicated the Bardia animals were Asian elephants (*Elephas maximas*), not a new or resurrected species.

The nasal bridge and the domes have been documented before,

The giant constricting snake reported from South America.

usually on old and large Asian elephants. In Peter Byrne's book *Tula Hatti: The Last Great Elephant*, he published video stills of the title character, a huge and undoubtedly aged beast which Byrne believed to be the world's largest living Asian elephant. Tula Hatti did indeed have domes on his head, although they were less prominent than on the Bardia specimens. There was no nasal bridge.

While individual features of the Bardia animals may appear on other elephants, there is no record of a population in which huge size, nasal bridges, and head domes are all common. A unique subspecies or race may indeed be emerging in this remote valley. One theory is the animals' isolation created a "genetic bottleneck" which brought forth recessive genes, resulting in a similarity to *E. hysudricus*.

Cryptozoologists point out that, whatever the classification of these beasts, they are proof a truly gigantic animal of very distinctive appearance could remain completely unknown to science until the 1990s. That is something for zoology to think about—mammoths or no mammoths.

Add to this episode the many other discoveries made in just the past decade, along with the mysteries still to be solved—from the jungles of the Congo to the forests of Tasmania, from the snows of the Himalayas to the depths of the oceans—and it becomes clear the zoological world is still a great museum of wonders, many yet to be unveiled. There is no shortage of animals yet to be discovered. If we have the wisdom and the will to preserve the Earth's last Edens, we will be rewarded by encounters with fascinating creatures that today are still waiting in the shadows.

RESOURCES

A BASIC LIBRARY OF CRYPTOZOOLOGY

There is far more to cryptozoology than books, but books are a very useful starting point. Given the profusion of questionably scientific works on the subject, those new to cryptozoology might profit from the advice of someone who has waded through most of the relevant literature. Accordingly, I offer here a basic reading list of the classics in the field.

Any list of cryptozoology books has to start with Dr. Bernard Heuvelmans' *On the Track of Unknown Animals*. Heuvelmans' collection of cryptozoological reports from every inhabited continent is invaluable, even if few of the animals he surveyed turn out to be real. (I personally believe Heuvelmans made one major error, namely his endorsement of de Loys' "ape" photograph from South America.) Look for the slightly revised and updated version published in 1995 by Kegan Paul International.

Almost as fundamental is Willy Ley's *Exotic Zoology*. This book, published by Viking Press in 1959, collected the cryptozoological material from Ley's several books of scientific oddities. While some of the information is outdated, the book remains a fascinating and highly readable collection of zoological discoveries and mysteries.

Four other books help complete a survey the field of cryptozoology. Dr. Karl Shuker's *The Lost Ark: New and Rediscovered Animals of the Twentieth Century* (HarperCollins, 1993) describes the entire variety of animal finds made in the previous ninety-three years. This book was updated and reissued in 2002 as *The New Zoo*, although that title is extremely hard to find.

Dr. Roy Mackal's *Searching for Hidden Animals* (Doubleday, 1980) brings some of the major mystery-animal cases up to the date of its publication. And, while it may seem very egotistical for a writer to nominate his own book, I honestly feel that my *Rumors* of Existence (Hancock House, 1995) belongs here. *Rumors* surveys the most recent discoveries, the possible-extinction cases, and the better-documented mystery animals in one compact package. Loren Coleman and Jerome Clark's well-written *Cryptozoology A to Z*

(Simon and Schuster, 1999) takes a somewhat similar approach, but accepts a broader range of data and focuses more on the unconfirmed animals. Accordingly, only a fraction of the material in the two volumes overlaps, so serious cryptozoologists will want Coleman's book handy as well.

There are several good books concerning large animals reported from the oceans. A collection of the best starts with Rupert T. Gould's *The Case for the Sea Serpent* (Phillip Allan, 1930). Writing when many witnesses to the "classic" cases were still alive, Gould presented a formidable brief in favor of the existence of one or more "monsters." Four decades later, Bernard Heuvelmans assembled all the known information on the subject in his massive *In the Wake of the Sea-Serpents* (Hill and Wang, 1968). Sifting over 300 "good" cases, Heuvelmans once again provided a great service, although his belief in up to seven large unknowns is difficult to credit. Richard Ellis brought the subject up to date with his *Monsters of the Sea* (Alfred Knopf, 1994), a skeptical but open-minded and thorough treatment of the subject throughout history.

Two books, both entitled *Bigfoot*, deserve to be read by all serious students of reported apelike creatures despite the passage of time since their publication. The first, by eminent primatologist John Napier (Berkley, 1972—there is also a 1975 edition), remains the best single volume ever written on the subject. It includes Asian as well as North American cases. Napier concluded that the yeti was a myth but sasquatch was probably real. Anthropologist Kenneth Wylie's book of the same name (Viking Press, 1980) provides a well-written skeptical counterpoint concerning sasquatch. Robert Pyle's *Where Bigfoot Walks* (Houghton Mifflin, 1995) puts the subject in cultural and geographical context, while John Bindernagel's 1998 work, *North America's Great Ape: the Sasquatch* (Courtenay, B.C., Canada: Beachcomber Books) offers a more current review by a qualified scientist. Henry Bauer's *The Enigma of Loch Ness* (the latest edition is from Johnston and Bacon Books, 1991) is probably the most balanced, readable book on Nessie. Dr. Karl Shuker's well-researched *Mystery Cats of the World* (Robert Hale, 1989) is the basic reference for those interested in that topic.

Well, those are my selections for the books making up the fundamental crypto-library. Some readers will undoubtedly suggest I've overlooked valuable works by John Green, Tony Healy and Paul Cropper, Grover Krantz, June O'Neill, Ivan Sanderson, Loren Coleman, Maurice Burton, Daniel Taylor-Ide, Daniel Cohen, and others. Perhaps I have, but one must start somewhere. Some of these books, along with many additional works, are reviewed in more depth in the next section.

REVIEWS: SELECTED BOOKS ON ZOOLOGY AND CRYPTOZOOLOGY

NOTE: This section is not meant to list every relevant book available, but I had had to be especially selective with Sasquatch books. Sasquatch is the subject of more books than all the rest of cryptozoology put together. It's impossible to include all those books here, even after throwing out the sensationalist works. Accordingly, this section includes only those books which, in the author's opinion, have a made a significant impact in some way.

Crowe, Philip Kingsland. 1967. ***The Empty Ark.*** New York: Charles Scribner's Sons. 281pp. This book recounts Crowe's travels around the world as an ambassador for the World Wildlife Fund. Everywhere Crowe went, he asked local experts about rare and strange creatures. As a result, there are tidbits in this book on cryptozoological subjects including giant anacondas, an unknown South American "bear," the survival of Przewalski's horse in the wild, the yeti, the almas, new species of bats, the Fiji petrel, the moa, the takahe, the thylacine, and the Iriomote cat. Crowe produced a first-rate book which is still valuable today.

Hunt, John. 1969. ***A World Full of Animals.*** New York: McKay. 378pp. Hunt, a conservationist and zookeeper, produced a now-dated but still highly enjoyable tour of the vertebrates. For cryptozoologists, there's a good discussion of how the tarpan was "re-created," plus reports and rumors of African bears, wild Przewalski's horses, and surviving thylacines. Most intriguingly, Hunt reported (without giving a source) that "Russian reports" indicated that not only was Steller's sea cow alive, but that a surviving population was under official protection.

Byrne, Peter. 1975. ***The Search for Bigfoot.*** Washington, D.C.: Acropolis Books. 263pp. Hunter and conservationist Byrne here recorded his unsuccessful efforts to find North America's reported ape. The greatest value of the book is not in the narrative, but the many illustrations, maps, and documents Byrne collected.

Green, John. 1980. ***On the Track of the Sasquatch***. Blaine, WA: Hancock House. 64pp. In this slender collection, Green offers the highlights of this first two decades of sasquatch-hunting. Whatever the explanation for sasquatch turns out to be, this is an interesting recap by one of the creature's most ardent pursuers.

Whitlock, Ralph. 1981. ***Birds at Risk***. Wiltshire, UK: Moonraker Press. 159pp. Whitlock compiled a large amount of data on the world's rarest birds, including a surprising number of species whose status (i.e., living or extinct) was unknown at the time of writing. This book also includes information on many recent discoveries.

Zhenxin, Yuan, and Huang Wanpo. 1981. ***Wild Man: China's Yeti***. London: Fortean Times. 23pp. This booklet is basically a collection of articles on China's reported unknown primate, with a helpful glossary of Chinese "monster" names and types. Includes analysis of hair and other evidence.

Blashford-Snell, John. 1983. ***Mysteries***. London: Bodley Head. 251pp. Blashford-Snell is an explorer who recounts ventures into the most remote parts of the world. He has an interest in cryptozoology and tells of his innovative but unsuccessful searches at Loch Ness and Loch Morar. He also discusses dinosaur legends in New Guinea, his encounters with outsized monitor lizards, and his meeting with "Oliver," the curious primate. Captivating adventure reading.

Shackley, Myra. 1983. ***Still Living?*** New York: Thames and Hudson. 192pp. Well-researched review of the evidence for mystery primates worldwide, with a special focus on Asia.

Guiler, Eric. 1985. ***Thylacine: The Tragedy of the Tasmanian Tiger***. Oxford, U.K.: Oxford University Press. Guiler, a zoologist who spent his professional life studying the thylacine and trying to determine if it still existed, tells the story of the animal and his quest for it in this book, which he followed with 1998's ***Tasmanian***

Tiger: A Lesson To Be Learnt (Perth, Australia: Abrolhos Publishing. 256pp.) In the second book, Guiler offers the reluctant conclusion that the thylacine is no longer alive.

Gaal, Arlene. 1986. ***Ogopogo***. Blaine, WA: Hancock House. 128pp. Gaal, who believes she has seen Ogopogo several times, is the closest thing this cryptid has to an official historian. This is a brief but lively account of the Canadian lake monster, with an excellent photo section and a table of sightings from 1860 to 1984.

Zarzynski, Joseph. 1986. ***Monster Wrecks of Loch Ness and Lake Champlain***. Wilton, NY: M-Z Information. 112pp. Unusual collection of short items about the two lakes in the title. The subjects include shipwrecks, archeology, history, and, of course, the search for "monsters."

Burton, John A., and Bruce Pearson. 1987. ***The Collins Guide to the Rare Mammals of the World***. Lexington, MA: The Stephen Greene Press. 240pp. Authoritative handbook of the world's rarest mammals, with copious color illustrations and range maps. Despite its date, this remains an invaluable reference. (It was reprinted in 1992.)

Mackal, Roy. 1987. ***A Living Dinosaur?*** Leiden, The Netherlands: E.J. Brill. 340pp. Mackal chronicles his arduous expeditions in search of reported unknown animals in the swampy Likouala region of the Congo. While the evidence reviewed in this book is limited to eyewitness reports and a single trail of strange footprints, Mackal believes there are as many as three large unclassified animals in this area. His most controversial belief is that one of these could be a small sauropod dinosaur.

Meurger, Michael, and Claude Gagnon. 1988. ***Lake Monster Traditions: A Cross-Cultural Analysis***. London: Fortean Tomes. 320pp. This unique scholarly work surveys monster reports and folklore from around the world and explores the common themes involved. While not entirely rejecting the possibility of "real" lake monsters, the authors trace the cultural roots of the subject in a way no one else has done.

Mountfort, Guy. 1988. ***Rare Birds of the World***. London: Collins. 256pp. Similar to Burton and Pearson's *Rare Mammals* (above), this is a wonderfully useful guide to the 1,000 most-threatened species of birds at the time of writing. Anyone with an interest in new, rediscovered, or vanishing birds simply must have this work handy at all times.

Rall, Kesar. 1988. ***Lore and Legend of the Yeti.*** Thamjel, Kathmandu, Nepal: Pilgrims Book House. 89pp. In this slender but valuable book, Rall assembled a collection of reports, lore, and legend concerning the variety of primates which reside either in the valleys of the high Himalayas or in the folklore of its people.

Bright, Michael. 1989. ***There Are Giants in the Sea***. London: Robson Books. 224pp. Bright, a biologist, offers an enjoyable survey of large and unusual sea creatures (the known and the unknown). He writes well and has obviously done his research, but the book is seriously weakened by the lack of references.

Coleman, Loren. 1989. ***Tom Slick and the Search for the Yeti***. Boston: Faber and Faber. 171pp. Coleman traces the yeti and sasquatch expeditions financed by oilman Tom Slick, which began in 1956 and ended with Slick's death in 1962. Reprinted in 2002 as ***Tom Slick: True Life Encounters in Cryptozoology***.

Day, David. 1989. ***Vanished Species***. New York: Gallery Books. 288pp. Superb artwork and well-researched text grace this large-format book on animals known or believed to have gone extinct within historical times. An entertaining and thought-provoking work.

Hutchison, Robert A. 1989. ***In the Tracks of the Yeti.*** London: MacDonald and Co. 285pp. Hutchison is an avid yeti-seeker who understands the people and cultures forming the backdrop to the yeti reports. He is positive the creature exists. On his expedition, Hutchison found several sets of what he thought were yeti tracks and examined a mummified "yeti foot" which he thought was very strange but which he fails to describe in detail. Most valuable if read in company with Daniel Taylor-Ide's similarly thorough but more skeptical work, ***Something Hidden Behind the Ranges*** (below). Taken together, these two books present most of what is known about the yeti mystery and introduce the reader to the cultural milieu in which it exists, providing vital information for evaluating reports of unknown primates.

Adams, Douglas, and Mark Cardawine. 1990. ***Last Chance to See***. New York: Harmony. 220pp. Writer Adams and zoologist Cardawine describe their travels to see the most endangered animals in the world, including the Komodo dragon and Yangtze river dolphin. A well-written, often humorous, and moving account.

Bergman, Charles. 1990. ***Wild Echoes***. Bothell, Washington: Alaska Northwest Books. 322pp. Bergman, an environmentalist, recounts his efforts in search of North America's most endangered animals, including the ivory-billed woodpecker, the Florida panther, and the black-footed ferret. Bergman managed the difficult trick of writing a very personal, often poignant, book which is also informative and highly educational.

Ciochon, Russell, with John Olsen and Jamie James. 1990. ***Other Origins: The Search for the Giant Ape in Human Prehistory.*** New York: Bantam. 262pp. This tale of scientific discovery presents what we know about Gigantopithecus and recounts expeditions to Vietnam in quest of more fossils. The authors dismiss the idea of surviving giant apes. Required reading for anyone interested in "unknown ape" questions.

Miller, Marc E. W. 1990. ***Chasing Legends.*** Stelle, IL: Adventurers Unlimited Press. 220pp. ***The Legends Continue***. 1998. Kempton, IL: Adventurers Unlimited Press. 229pp. Miller, a psychologist, is a dedicated amateur explorer and cryptozoologist.

These books recount his travels in search of the yeti, mokele-mbe-mbe, giant anacondas, African ape-men, and other cryptids. Miller reports two exhilarating moments: having his boat lifted by a large unidentified aquatic animal on a Venezuelan river, and photographing the strange elephants of Bardia. The writing could be more polished, and Miller makes some errors when he strays from his first-hand narrative, but the reader will certainly gain an appreciation for the challenges of cryptozoology.

Salvadori, Francesco. 1990. ***Rare Animals of the World.*** New York: Mallard Press. 192pp. This tour of the world's rarest creatures, illustrated with Piero Cozzaglio's paintings, includes a number of animals declared extinct but found again. These include, among other species, the Seychelles owl, the Persian fallow deer, and the bridled nail-tailed wallaby.

Stap, Don. 1990. ***A Parrot Without a Name***. New York: Alfred A. Knopf. 239pp. Enthralling account of adventures with ornithologists John O'Neill and the late Ted Parker in search of rare and new species of birds in Peru.

Cone, Joseph. 1991. ***Fire Under the Sea.*** New York: William Morrow. 285pp. An exciting account of the discovery of seafloor volcanic vents and the new and strange life forms they support. Well illustrated and referenced.

Ellis, Richard, and John E. McCosker. 1991. ***Great White Shark***. New York: HarperCollins. 270pp. Ellis, along with marine biologist McCosker, has collected every known fact about the sea's most feared predator in this thoroughly illustrated, referenced, and readable book. The authors discuss the truth about the largest documented white shark ever caught (around twenty-two feet long). Another chapter is devoted to the claims that *Carcharadon megalodon* might still be alive, something the authors doubt. Fascinating reading.

Thomson, Keith S. 1991. ***Living Fossil: The Story of the***

Coelacanth. New York: W.W. Norton. Dr. Thompson covers nearly everything about the coelacanth, including the question of distribution, in this authoritative book. The only topic he does not address is the hint of a New World population.

Dietz, Tim. 1992. ***The Call of the Siren***. Golden, CO: Fulcrum Publishing. This is a well-written and affectionate book on the manatee and its relatives, paying special attention to the Florida manatee. A good discussion of Steller's sea cow and its possible survival is included.

Gould, Charles. 1992. ***Mythical Monsters***. London: Studio Editions. 407pp. In this reprint of a book published in 1884, the reader will find many oddities of natural history (some, not surprisingly, have been explained or disproved over the subsequent 120 years). Of most interest is the sea serpent section. Gould believed one or more species of unknown animals were involved in sea serpent reports. In this book, he included many reports made before and around his time, including cases where Gould talked to the witnesses himself.

Krantz, Grover. 1992. ***Big Footprints: A Scientific Inquiry into the Reality of Sasquatch***. Boulder, CO: Johnson Books. 300pp. In this book, anthropologist Krantz sifted through sasquatch reports and tried to deduce the animal's lifestyle and characteristics based on what he considered the most reliable information. It's a good effort, although necessarily speculative. Krantz weakened his argument by accepting partial evidence in two cases where some faking was involved, but this is still an important work on the subject.

Walters, Mark Jerome. 1992. ***A Shadow and a Song: The Struggle to Save an Endangered Species***. Post Mills, VT: Chelsea Green Publishing Co. A poignant chronicle of the decline and extinction of the dusky seaside sparrow. Walters also explores the controversial and still somewhat mysterious events that brought an end to the efforts to save the dusky.

Alderton, David. 1993. ***Wild Cats of the World***. New York: Facts on File. A thorough compendium of cat lore and cat taxonomy. All known species are here, along with a good discussion of the peculiar hybrid called the Kellas cat. Alderton gives a detailed account of the onza, which he boldly lists as a species in its own right, a hypothesis since disproved.

Brakefield, Tom. 1993. ***Kingdom of Might: The World's Big Cats***. Stillwater, MN: Voyageur Press. A fascinating and sumptuously illustrated tour of the great cats. Brakefield also discusses feline hybrids and the color variations of each species. Cryptozoologists will be disappointed by what's missing: there is nothing on the marozi (spotted lion) or other problematical cats.

Francis, Di. 1993. ***The Beast of Exmoor***. London: Jonathan Cape. Francis' first book, 1983's *Cat Country*, advanced the highly controversial theory that Britain had an indigenous unclassified big cat. In this book, she broadens her scope to include the Kellas cat and a bizarre carcass from Dufftown, Scotland, that most authorities consider a freak. Francis' tenacious support of the indigenous big cat (a difficult theory to swallow, even for other cryptozoologists) will turn off some readers, but there are certainly a lot of interesting reports here.

Hunter, Don, with René Dahinden. 1993. ***Sasquatch/Bigfoot***. Buffalo, NY: Firefly Books. 205pp. Chronicle of the title subject, focusing on Dahinden's long hunt for sasquatch. The book is sometimes one-sided, but it does include descriptions of exposed hoaxes and is worth reading for students of this topic.

Jacobs, Lewis. 1993. ***Quest for the African Dinosaurs***. New York: Villard Books. 314pp. This book mainly recounts Dr. Jacobs' pioneering fieldwork on African dinosaur fossils. Jacobs spends one chapter on allegations of living dinosaurs in the Congo. His tone is overly derisive, but his doubts are logically based on the lack of evidence and the fact that the area involved has changed since the Mesozoic and is not a "lost world" where relict species are likely.

McClung, Robert M. 1993. ***Lost Wild America***. Hamden, CN: Linnet Books. This is a first-rate work covering the history of extinctions and endangered species in and around the United States. A wealth of data includes profiles of lost, rediscovered, and rare animals of all types, woven into a very readable narrative and illustrated with drawings. Thorough bibliography.

Moffett, Mark W. 1993. ***The High Frontier***. Cambridge, MA: Harvard U. Press. Explores Earth's least-known land environment —the tropical rainforest canopy. Moffett also describes Terry Erwin's studies of canopy arthropods and his predictions of the total number of species yet to be found. Good bibliography and excellent photographs.

Nugent, Rory. 1993. ***Drums Along the Congo.*** New York: Houghton Mifflin. This is a chronicle of Nugent's adventures in search of the alleged living dinosaur of the Congo, which the author believes he spotted. While Nugent includes photographs, the object shown is much too distant for identification.

Sylvestre, Jean-Pierre. 1993. ***Dolphins and Porpoises: A Worldwide Guide.*** New York: Sterling. 159pp. Thorough treatment of the world's small cetaceans. Included are such cryptozoological tidbits as a possible dwarf spinner dolphin, anomalous killer whales, and species whose identity is in dispute.

Bonner, Nigel. 1994. ***Seals and Sea Lions of the World.*** New York: Facts on File. 224pp. Bonner has written an authoritative guide to the known pinnipeds. Items of special interest include two species thought extinct but rediscovered, the Juan Fernandez and Guadeloupe fur seals, and two which may be extinct, the Japanese sea lion and the Caribbean monk seal.

Ellis, Richard. 1994. ***Monsters of the Sea***. New York: Alfred A. Knopf. 429pp. Richard Ellis covers "sea monsters" real, fictitious, and undetermined. The book is thoroughly referenced, expertly written, and generally fascinating. Ellis believes most sea monster sightings are mistakes, often involving giant squid. He accepts, though, that the repeated sightings off Gloucester in 1817 defy easy explanation. Ellis also surveys the modern reports of "Cadborosaurus" from the coastal Pacific. He concludes, "If I were a betting man, I would bet against it. (I would, however, like nothing better than to lose the bet.)" Ellis reviews the Saint Augustine, Florida carcass from 1896 with great thoroughness and concludes it involved a genuine giant octopus. Ellis has written an enjoyable and important book, whether all his conclusions are correct or not. The extensive bibliography shows the kind of research only the best writers put into their work.

Healy, Tony, and Paul Cropper. 1994. ***Out of the Shadows: Mystery Animals of Australia***. Chippendale, Australia: Ironbark. 200pp. A balanced and thoroughly researched study of the region's six most interesting cryptozoological puzzles. These include the thylacine's survival on Tasmania, the same animal's possible survival in Australia, alien big cats, the Queensland "tiger," the apelike "Yowie," and the legendary bunyip. The authors assign them credibility in about that order, although they actually build the strongest case for the introduced American puma. Healy and Cropper lose scientific credibility only in the final chapter, where they discuss "psychic animals" (apparitions) as an explanation for some of the reports. I fear zoologists will discard this otherwise valuable book when they get to that point.

Humphreys, Charles R. 1994. ***Panthers of the Coastal Plain.*** Wilmington, NC: The Fig Leaf Press. 200pp. Reports of supposedly-extinct Eastern cougars in North Carolina. Humphreys may be too quick to accept all the witnesses' statements, but he has assembled a valuable database of 160-plus accounts.

Pianka, Eric R. 1994. ***The Lizard Man Speaks.*** Austin, TX: University of Texas Press. 179pp. The entertaining and informative adventures of a herpetologist on his collecting trips. The reader will learn a great deal about lizards, and Pianka includes the stories of his discovery and classification of several new species.

Thomas, Elizabeth Marshall. 1994. ***The Tribe of Tiger.*** New York: Simon and Schuster. 240pp. Thomas' fascinating, if sometimes speculative, look at the world's variety of cats contains information about the "extinct" Eastern cougar, including Thomas' personal sighting.

Wade, Nicholas, with Cornelia Dean and William A. Dicke (Eds.). 1994. ***The New York Times Book of Science Literacy, Volume II: The Environment from your Backyard to the Ocean Floor.*** 480pp. This imposingly titled book has several articles of interest. These include "A Cornucopia of Life," describing the diversity of the seafloor and how little we know of its denizens: "The Aye-Aye," an update on the bizarre little primate once feared extinct: and "To Conserve or Catalog Rare Species," concerning the discovery of Somalia's boubou shrike and the controversial decision to release the type specimen back into the wild.

Weiner, Jonathan. 1994. ***The Beak of the Finch.*** New York: Alfred A. Knopf. 332pp. Weiner's award-winning study on the finches of the Galapagos Islands. The book includes valuable explanations about DNA, hybrids, natural selection, and the dynamics of small, isolated animal populations.

Ackerman, Diane. 1995. ***Rarest of the Rare.*** New York: Random House. 184pp. In this beautifully written travelogue, Ackerman goes in search of some of the rarest animals in the world. Of special interest is her voyage to the island of Torishima to see the short-tailed albatross (*Diomedea albatrus*), which was declared extinct in 1949 but clung to an extremely precarious existence on this speck of volcanic rock. Ackerman displays a poetic, almost magical writing style that brings the reader irresistibly along on her adventures.

Ballard, Robert D. 1995. ***Explorations***. New York: Hyperion. 407pp. In this fascinating account of Ballard's undersea discoveries, the explorer and geographer tells about the finding of a totally new ecosystem—the deep-sea hydrothermal vent colonies. Ballard recounts how the first report was received with disbelief by biologists, and how the myriad of animals discovered at these vents has contributed to our understanding of the origin and development of life.

Bass, Rick. 1995. ***The Lost Grizzlies.*** New York: Houghton Mifflin. 240pp. One of two recent books on the possible survival of the grizzly bear (*Ursus arctos horribilis*) in the state of Colorado, where these huge predators were supposedly exterminated in the 1950s. Bass, a first-rate wilderness writer, believes he personally got a brief glimpse of one of these wary bruins.

Cardawine, Mark. 1995. ***Whales, Dolphins and Porpoises: The Visual Guide to all the World's Cetaceans.*** London: Doris Kindersley. 256pp. Well written and superbly illustrated guide including two cryptic species. Martin Camm's illustrations include a speculative drawing of *Mesoplodon* (or *Indopacetus*) *pacificus*, at the time of publication known only from two skulls and a few unconfirmed sightings, and one of "Species A," the widely reported and still mysterious beaked whale of the Eastern Tropical Pacific.

Carmony, Neil. 1995. ***Onza!*** Silver City, NM: High-Lonesome Books. 204pp. In this enjoyable book, Carmony chronicles a century of interest in Mexico's legendary big cat. Carmony concludes, with obvious regret, that the onza specimens collected so far show the cat is just an odd puma.

Corliss, William. Corliss' organization, the Sourcebook Project, continues to publish valuable collections of zoological and cryptozoological material. Two examples are ***Biological Anomalies: Mammals I*** (1995) and ***Mammals II*** (1996). A wealth of data on thylacines, Nandi bears, sea serpents, mystery whales, and many other subjects is found here, all of it well-referenced. The Sourcebook Project also publishes the newsletter *Science Frontiers*. For a booklist, write to Sourcebook Project, P.O. Box 107, Glen Arm, MD, 21057, U.S.A, or visit http://www.science-frontiers.com/index.htm.

Flannery, Tim. 1995. ***Mammals of New Guinea***. New York: Cornell University Press. 538pp. Dr. Flannery collects all the known mammals of the New Guinea region, including many recent finds. Of special interest is Flannery's account of his search for the newest tree kangaroo, the dingiso. This sizable mammal was not confirmed by Western science until Flannery collected the first specimens in 1994. Flannery also includes sections describing the region's geography, ecology, and paleontology.

Heuvelmans, Bernard. 1995. ***On the Track of Unknown Animals.*** London: Kegan Paul International. 677pp. This is an update to the founding work of cryptozoology, first published in 1955. The updating unfortunately consists only of a thirteen-page preface to the original text, plus some new illustrations. Still, the amount of material here is impressive: the bibliography fills twenty-six pages. The book includes alleged animals of South America, Africa, Asia, and Australasia, ranging from the famous (like the yeti) to little-known ones like Australia's dinosaur-like *gauarge*.

Heuvelmans does a great service in his examination of matters like the "Nandi bear" of Africa. Time and again, he patiently untangles a welter of contradictory reports, explains most of them, and offers candidates for the ones he feels remain unexplained. I can't help wishing Heuvelmans had rewritten his original text and updated it throughout. Nevertheless, this is the most influential single volume on land-dwelling cryptids ever published in English.

Payne, Roger. 1995. ***Among Whales.*** New York: Scribner's. 431pp. The only item of cryptozoological interest is a discussion of how many species of right whale there are, but Dr. Payne provides a superb introduction to cetaceans. For researchers who want a basic background in cetacean science, I recommend this along with Richard Connor and Dawn Peterson's ***The Lives of Whales and Dolphins*** (1994. New York: Henry Holt) as the first books to read.

Peterson, David. 1995. ***Ghost Grizzlies.*** New York: Henry Holt. 296pp. This enthralling zoological detective story takes readers on the trail of a possible relict population of grizzly bears in

Colorado's San Juan Mountains. While the proof is not yet final, the reader will learn much about bears, conservation, and animal tracking in this book.

Pyle, Robert Michael. 1995. ***Where Bigfoot Walks.*** Boston: Houghton Mifflin. 338pp. A unique exploration of the myth and reality surrounding sasquatch. Pyle, an ecologist, does not attempt to prove or disprove the animal's existence. He goes exploring with an open mind and, while providing insights on conservation, spotted owls, and dedicated sasquatch hunters, makes a strong argument that such primates could find food and remain hidden in the U.S. Northwest.

Shuker, Karl P. N. 1995. ***Dragons: A Natural History***. New York: Barnes and Noble. 120pp. An enjoyable romp through dragons of myth, fable, and legend, mixed with cryptozoological reports of creatures which may have inspired dragon stories or been inspired by them.

Shuker, Karl P. N. 1995. ***In Search of Prehistoric Survivors***. London: Blandford. 192pp. In this fascinating work of cryptozoology, Dr. Shuker has gathered information on reported animals around the world. His tone is sometimes more credulous than I would like: it is, for instance, almost impossible to credit that the world's largest flying bird might be living undiscovered in the Ohio Valley of the United States. Nevertheless, this is an indispensable volume for everyone interested in undiscovered animals. Shuker has done a massive amount of research and shown that in some cases (such as that of the Australian marsupial tiger or *yarri*) the evidence is much stronger than is usually thought.

Strahan, Ronald (editor). 1995. ***Mammals of Australia***. Washington, D.C.: Smithsonian Institution Press. 756pp. This massive reference work is indispensable to those seeking crypto-marsupials. It provides a good portrait of the "extinct" thylacine and offers an authoritative compendium of the animals recently discovered or rediscovered in Australia.

Taylor-Ide, Daniel. 1995. ***Something Hidden Behind the Ranges.*** San Francisco: Mercury House. 298pp. Taylor-Ide, born in the shadow of the Himalayas, here recounts his personal search for the Yeti. He never did find it, and eventually reached the conclusion that no such primate existed. Along the way, however, he made important discoveries about bears, conservation, and the unique place the yeti holds in the human mind and spirit.

Blashford-Snell, John, and Rula Lenska. 1996. ***Mammoth Hunt***. London: HarperCollins. 263pp. Explorer Blashford-Snell recounts his successful quest for the rumored "monster elephants" of Bardia. The book combines an exciting travelogue with a good study of elephants.

Ellis, Richard. 1996. ***Deep Atlantic: Life, Death, and Exploration in the Abyss.*** New York: Alfred A. Knopf. 395pp. Few authors could write a first-class book solely about the environment deep under the Atlantic Ocean, but Ellis has managed it. Meticulously researched as usual (the bibliography covers 47 pages), this work includes many tidbits for the cryptozoologist. Examples are Ellis' favorite creatures, the giant squids; the weird fish reported by pioneer undersea explorer William Beebe; and the now-famous hydrothermal vent colonies with their otherworldly inhabitants. Also valuable is a discussion of the six-foot larva once known as *Leptocephalus giganteus*.

Quammen, David. 1996. ***The Song of the Dodo***. New York: Scribner's. 702pp. The zoological writings of David Quammen are best dealt with as a group. Most of Quammen's books collect his articles from ***Outside*** and other publications, while ***The Song of the Dodo*** is an original work of great significance.

In 1985's ***Natural Acts***, Quammen includes two essays of cryptozoological significance. "Avatars of the Soul in Malaya" concerns discoveries in the world of moths and butterflies, including some very odd new species. In "Rumors of a Snake," Quammen collects the stories of a giant anaconda in South America. Of the famous claim by Major Percy Fawcett to have shot a snake sixty-two feet long, Quammen observes, "It might all be true but most likely it isn't."

In The ***Flight of the Iguana*** (1988), Quammen includes an essay entitled, "Stranger than Truth: Cryptozoology and the Romantic Imagination." Quammen holds the common view that "cryptozoology is biased toward large unknown animals." He seems skeptical about the existence of such animals but thinks well of the scientists involved and describes the journal ***Cryptozoology*** as "intriguing, diverse, and mainly quite sane."

The Song of the Dodo is subtitled ***Island Biogeography in an Age of Extinction***. In studying what happens to "island" populations—how they change, how many individuals they need to remain viable, and how they become extinct—Quammen deals not only with literal islands but with isolated areas on larger land masses. Two sections cover the 1986 discovery of the golden bamboo lemur (*Hapalemur aereus*) and the disappearance of the thylacine.

In Quammen's ***The Boilerplate Rhino*** (2000), the essay "Limelight" covers the media circus concerning the 1994 sighting of two octopi of unknown species. (The seemingly confused cephalopods, both male, were filmed while trying to mate. See Section I of the present book.)

Rowe, Noel. 1996. ***The Pictorial Guide to the Living Primates***. East Hampton, NY: Pogonias Press. 263pp. This handbook offers brief, clear descriptions of all known primate species, good illustrations, and range maps, plus an explanation of the taxonomic issues concerning each species (and it's surprising how many species' taxonomy remains unsettled). Accurate but not too technical for the lay reader, this is a first-rate addition to anyone's reference shelf.

Smith, Malcolm. 1996. ***Bunyips And Bigfoots: In Search of Australia's Mystery Animals***. Smith, a zoologist, has written a terrific book: enjoyable, fact-filled, and skeptical but open-minded in tone. He doubts the existence of both the title animals, but concludes there are some real mysteries to be solved. Smith believes Australia probably does house imported big cats, though he can't imagine how they arrived, and is also impressed by some "sea serpent" reports from Australian waters.

Tischendorf, Jay, and Steven J. Ropski (editors). 1996. ***Proceedings of the Eastern Cougar Conference, 1994***. Fort Collins, Colorado: AERIE. 245pp. Collected papers on all aspects of the Easter cougar question. An invaluable resource.

Van Dover, Cindy Lee. 1996. ***The Octopus's Garden: Hydrothermal Vents and Other Mysteries of the Deep Sea.*** New York: Helix Books. 183pp. In this fascinating account of probes into the ocean depths, Van Dover shows the reader the variety of bizarre environments recently discovered under the sea and their even more bizarre inhabitants: the famous tubeworms, "spaghetti worms," and even a shrimp named *Chorocaris vandoverae* after the author.

Broad, William J. 1997. ***The Universe Below.*** New York: Simon and Schuster. 432pp. Broad's book on the recent discoveries under the sea covers oceanography and other matters not directly related to marine zoology, but there's enough about new denizens of the deep to keep any cryptozoologist turning the pages. Broad includes a comment from Dr. Robison of the Monterey Bay Aquarium Research Institute, who says we have so far missed classifying at least one third of the large species in the sea. "We may find that's conservative," he adds. "It could be half."

Campbell, Steuart. 1997. ***The Loch Ness Monster: The Evidence***. Amherst, New York: Prometheus Books. 128pp. Steuart surveys all the evidence concerning Loch Ness and finds it wanting. His skepticism seems dogmatic, but many of his explanations are reasonable.

McClung, Robert M. 1997. ***Last of the Wild***. Hamden, CN: Linnet Books. 291pp. Survey of extinct and endangered animals around the world. From the saola to the thylacine (which McClung believes was alive in 1982), there are countless items of interest in this well-written book. The extensive bibliography is a bonus.

Murphy, John C., and Robert W. Henderson. 1997. ***Tales of Giant Snakes: A Historical Natural History of Anacondas and***

Pythons. Malabar, Florida: Krieger Publishing. 221pp. The authors here recount all that is known about the history, habits, and sizes of the world's largest snakes. They are skeptical of the idea of an entirely unknown giant in the Amazon basin. Murphy and Henderson conclude that three species (the anaconda and the two largest pythons) "appear to approach, and possibly slightly exceed 30 feet on occasion," although they include some credible-sounding reports of what may have been larger specimens. Mandatory reading for all those interested in the subject of giant snakes.

O'Hanlon, Redmon. 1997. ***No Mercy: A Journey to the Heart of the Congo.*** New York: Alfred A. Knopf. 480pp. In this travelogue, O'Hanlon undertakes a serious (yet sometimes humorously recounted) exploration of the Lake Tele region of the Congo, supposed home of a giant reptile. O'Hanlon never saw the creature, but collected some fascinating background on the area. He also describes a reported animal called the *Yombe*, which may be an unclassified ape.

Shuker, Karl. 1997. ***From Flying Toads to Snakes With Wings.*** St. Paul, Minnesota: Llewellyn Publications. 222pp. Based on cryptozoology articles Dr. Shuker published from 1988 to 1997. The most fascinating item is an account of a titanic jellyfish (with estimated 200-foot tentacles) impaled on the bow of a steamship in 1973. Unfortunately, no one has printed the analysis which supposedly confirmed this identity or the names of the experts who performed it. Dr. Shuker throws in some irrelevant items, such as "ghost dogs," but this collection is so interesting that the reader should be willing to forgive a few such flights of fancy.

Turner, Alan. 1997. ***The Big Cats and their fossil relatives.*** New York: Columbia University Press. 234pp. Turner discusses the current big cats and all known fossil forms, putting his subjects in context. Turner does not address any cryptozoological subjects, but provides a great deal of knowledge useful in evaluating "unknown cat" claims. The illustrations are superb.

Wade, Nicholas (editor). 1997. ***The Science Times Book of Fish.*** New York: Lyons Press. 288pp. This collection of *New York Times* articles on aquatic life offers several items of interest to the cryptozoologist. They include articles on our favorite fish, the coelacanth: the many bizarre new species brought up from the Amazon: and new freshwater fishes from Brazil. This last article introduces biologist Michael Goulding, who has described 400 new species.

Bindernagel, John. 1998. ***North America's Great Ape: the Sasquatch***. Courtenay, B.C., Canada: Beachcomber Books. 270pp. Rather than argue sasquatch's existence, Bindernagel essentially accepts that as a given and instead focuses on analyzing behavior, anatomy, and other characteristics of the animal. The result is intriguing, but necessarily speculative. Even the most ardent sasquatch proponents accept there are many hoaxes and mistakes, and there is no way Bindernagel can be certain all the evidence on which his theories rest is genuine.

Dixon, Dougal. 1998. ***After Man: A Zoology of the Future***. New York: St. Martin's Griffin. 124pp. While this speculative look at what might have evolved 50 million years after the Age of Man has no direct connection to cryptozoology, it's a splendid introduction to how animal evolution works and how creatures adapt to fill niches in their environments.

Eggleton, Bob, and Nigel Suckling. 1998. ***The Book of Sea Monsters.*** Woodstock, NY: Overlook Press. 112pp. This unique book is mainly a showcase for Eggleton's paintings of sea and lake monsters, real, reported, and mythical. Suckling's text is forgettable, but Eggleton's visions of Nessie, plesiosaurs, Ogopogo, etc., are beautiful works of talent and imagination.

Ellis, Richard. 1998. ***The Search for the Giant Squid***. New York: Lyons Press. 322pp. Ellis has searched exhaustively through the literature and lore of the squid, and the result is a highly readable and informative work. Looking at the favorite question of how big *Architeuthis* gets, he concludes, "we don't know." As always with an Ellis book, the bibliography is worth the price of the entire work: this one is 38 pages. A superb contribution.

Flannery, Tim. 1998. ***Throwim Way Leg***. New York: Atlantic Monthly Press. 326pp. Dr. Flannery has written a captivating book about his adventures seeking new species in New Guinea. Flannery has described two new species of kangaroos and over twenty other species and subspecies of mammals. This book is part zoology lesson and part travelogue. Flannery makes it very clear New Guinea still has large tracts of untrodden and almost inaccessible terrain where there are doubtless more species to look for. Must reading for those interested in New Guinea's fauna or in the challenges of penetrating Earth's last unknown regions for the sake of zoology.

Kirk, John. 1998. ***In the Domain of the Lake Monsters: The Search for Denizens of the Deep***. Toronto: Key Porter Books. Kirk, President of the British Columbia Scientific Cryptozoology Club, recounts his own search for "Ogopogo," the alleged inhabitant of Lake Okanagan, and discusses other "lake monsters" worldwide. This book indicates that Lake Okanagan is actually more likely than more famous Loch Ness to house an unknown animal. Kirk has created a very readable introduction to the lake monster world, albeit one based mainly on anecdotal evidence.

Parker, Gerry. 1998. ***The Eastern Panther: Mystery Cat of the Appalachians***. Halifax, Nova Scotia, Canada: Nimbus Publishing. 210pp. A history of the Eastern panther controversy, with detailed records of the last proven killings and an even-handed examination of the continuing debate over the animal's survival. Parker, a retired wildlife biologist, finds the evidence to be inconclusive. Still, he writes, "I remain encouraged that the panther may again be among us."

Steene, Roger. 1998. ***Coral Seas***. Buffalo, New York: Firefly Books. 272pp. This stunning photographic tour of coral reefs and other marine landscapes includes photographs of the Mimic

Octopus (see Section I) and several undescribed marine invertebrates. Also includes a photo of an albino humpback whale.

Allen, Thomas B. 1999. ***The Shark Almanac***. New York: Lyons Press. 264pp. Thorough introduction to the world of sharks and their relatives, including recent discoveries like Megamouth and rediscoveries like the Borneo River Shark. The same author produced 1996's ***Shadows in the Sea: The Sharks, Skates, and Rays*** (New York: Lyons and Burford), an update of one of the classic works in this field and also a very useful reference.

Coleman, Loren, and Patrick Huyghe. 1999. ***The Field Guide to Bigfoot, Yeti, and Other Mystery Primates Worldwide.*** New York: Avon. 207pp. Coleman and Huyghe devised a classification system in which over fifty reported unknown primates are sorted into nine groups. The authors do *not* claim that each group represents a valid animal. This book is useful for studying cryptid-primate reports because it lets the researcher put them in context with reports similar in description and/or location. The authors believe the top prospect for discovering a real animal is the orang-pendek of Sumatra. Even if most of the types discussed here don't really exist (I think at least two of their nine categories, concerning aquatic primates and "true giants" fifteen to eighteen feet tall, are surely invalid), any cryptozoologist trying to understand the subject of unclassified large primates will want this book. A good bibliography and a list of contacts around the world add to its value.

Coleman, Loren, and Jerome Clark. 1999. ***Cryptozoology A to Z.*** New York: Simon and Schuster. 267pp. For a book whose subtitle bills it as an "Encyclopedia," this is shorter than one might expect, but there's plenty of good material in here. The authors cover the major animals, publications, and personalities in cryptozoology, making this a first-rate introduction for the newcomer or a handy sourcebook for the experienced researcher. Two quibbles: I would have liked to see the references broken down by topic, and the wording in the entry on *C. megalodon* gives the impression I believe this extinct monster shark survives, when that isn't the case (Coleman has apologized for not being clearer). The authors necessarily had to leave a lot of mysteries out, but the important ones are all here, and the writing is clear and balanced. The bottom line is this book is a "must-have" addition to the cryptozoologist's library.

"Cryptozoological Society of London." 1999. ***A Natural History of the Unnatural World***. New York: St. Martin's Press. An elaborate spoof, mixing zoology, cryptozoology, myth, and outright fiction, all presented in a mock "scientific" tone. I list this book here only to warn that it's for entertainment value only. (The "Society" does not exist, and some cryptozoologists fear this book will deal a serious blow to the credibility they're trying so hard to acquire.)

Fuller, Errol. 1999. ***The Great Auk***. New York: Harry N. Abrams. 420pp. Beautifully written, sumptuously illustrated homage to the Great Auk (*Alca impennis),* one of the world's most famous extinct birds. Fuller believes a few birds probably survived the species' presumed demise in 1844, but the scattered sightings after that date provide no real hope the bird still lives.

Garbutt, Nick. 1999. ***Mammals of Madagascar***. New Haven, CT: Yale University Press. 320pp. Thorough, well-illustrated guidebook which includes everything from taxonomy to viewing opportunities for every known species of living mammal on the island. Also covers extinct species, but does not mention any of the reports concerning "extinct" mammals which might possibly survive.

Hall, Dennis. 1999. ***Champ Quest.*** Jericho, VT: Essence of Vermont. 90pp. Unusual collection of sighting reports, speculation, notes pages, and calendars of the best days to look for "Champ" (it depends on the phases of the moon). Hall's claims of spectacular close-up sightings of a type reported by no one else are bound to raise skeptical eyebrows.

Krantz, Grover. 1999. ***Bigfoot Sasquatch Evidence***. Blaine, WA: Hancock House. 348pp. Revised and updated version of the author's *Big Footprints* (above). Krantz vigorously defends the reality of the Patterson-Gimlin film and some of the trackways he's examined. Krantz is more cautious than some cryptozoologists when discussing worldwide reports of such creatures: he sees no reason to accept the yeti or to assume there's more than one species of unknown primate. Krantz added changes at the back of the book instead of rewriting the text, which makes the book a little hard to follow. It's still worth reading, though, as it summarizes the best arguments on the pro-sasquatch side.

Lowman, Margaret. 1999. ***Life in the Treetops***. New Haven, CT: Yale University Press. 219pp. Lowman, a biologist, recounts her pioneering work in exploring forest canopy environments on four continents. In the course of telling about her experiences, she provides information on forest ecology, botany, and the unusual creatures of the upper story.

Martin, David, and Alastair Boyd. 1999. ***Nessie: The Surgeon's Photograph Exposed***. Thorne Printing: 100pp. This small self-published book makes a strong argument (though others still dispute it) that the most famous Loch Ness photo was a fake.

Nowak, Ronald M. 1999. ***Walker's Primates of the World***. Baltimore: Johns Hopkins. 224pp. Fact-packed reference from one of the world's leading authorities, this book provides information on 282 living and recently extinct species. Nowak is also the author of the broader reference book, ***Walker's Mammals of the World***, which is unanimously considered the Bible of mammology. *Mammals* also came out in its 6th edition in 1999. (Johns Hopkins, 1248pp.)

O'Neill, June P. 1999. ***The Great New England Sea Serpent***. Camden, Maine: Down East Books. 256pp. This is a marvelous little book—thoroughly researched, enjoyable, and open-minded—about the great sea creature whose appearances off Cape Ann in the early 1800s captivated all of New England. O'Neill shows how sightings of a similar creature far predated the famous flurry of activity in 1817, and have continued, albeit much less frequently, into modern times. One of the best marine cryptid books ever written.

Purcell, Rosamond Wolff. 1999. ***Swift as a Shadow***. Boston: Houghton Mifflin. 160pp. A haunting collection of superb photographs of museum specimens of endangered and extinct animals. Specimens depicted include the thylacine, passenger pigeon, quagga, and Javan tiger. A touching, memorable book.

Shuker, Karl. 1999. ***Mysteries of Planet Earth***. Carlton Books. 192pp. Shuker again mixes cryptozoology with unrelated "phenomena." Whatever one thinks of apparitions and ancient engineering, there's enough well-written cryptozoology in here to make the book worth having. Shuker's interests range from animals with odd coloration to suspect hybrids to lake monsters. Shuker is good at turning up overlooked or little-known cryptid reports. We all know about Nessie, but the "whale-fish" of Lake Myllesjon in Sweden will be new to practically everyone. Paintings by Bill Rebsamen add to the book's appeal.

Wade, Nicholas (editor). 1999. ***The Science Times Book of Mammals***. New York: Lyons Press. 288pp. This entry in the valuable *New York Times* series includes 47 articles on mammals. Of special interest are pieces on David Oren's search for the reported ground sloth of the Amazon and the rediscovery of the woolly flying squirrel.

Cokinos, Christopher. 2000. ***Hope is the Thing With Feathers: A Personal Chronicle of Vanished Birds***. Los Angeles: J.P. Tarcher. 352pp. Eloquent story of Cokinos' investigation of the disappearance of six North American birds, including the presumed extinction of the ivory-billed woodpecker. Cokinos recounts sighting reports of ivory-bills as recent as 1999.

Messner, Reinhold (translated by Peter Constantine). 2000 (English-language edition). ***My Quest for the Yeti***. New York: St.

Martin's Press. 169pp. This is really two books: one, an adventure tale by a great mountaineer; the other, a confusing attempt to show the yeti is a large and very strange type of brown bear. Whatever one thinks of the yeti, the bear hypothesis seems a poor fit. Messner describes yetis routinely walking and running on two feet and includes photographs of an ordinary-looking brown bear he claims is the mysterious bipedal *chemo*, or yeti.

Tyson, Peter. 2000. ***The Eighth Continent: Life, Death, and Rediscovery in the Lost World of Madagascar***. New York: William Morrow. 374pp. This excellent book explores Madagascar, an island of zoological discovery and mystery. Tyson takes the reader to meet such pioneers as herpetologist Chris Raxworthy, who has a backlog of 150 new species to describe, and primatologist Patricia Wright, discoverer of a new lemur. Along the way, he describes the biodiversity of this endangered land and probes reports indicating two species of Madagascar's supposedly extinct megafauna—a pygmy hippo and a giant lemur—just might still exist.

Vrba, Elisabeth S., and George B. Schaller (editors). 2000. ***Antelopes, Deer, and Relatives: Fossil Record, Behavioral Ecology, Systematics, and Conservation***. New Haven, CT: Yale University Press. 341pp. Scholarly collection of papers and essays on all aspects of the title subject. The book includes two papers covering the new and rediscovered species in Southeast Asia.

Weinberg, Samantha. 2000. ***A Fish Caught in Time***. New York: HarperCollins. 220pp. Superbly written history of the coelacanth and the people who have pursued it. Less technical than Keith Thomson's book (above), this volume is filled with details about the effect this unique fish has had on those caught in its prehistoric aura. Weinberg includes a thorough account of the discovery of the second population in Indonesia, along with evidence for other populations.

Flannery, Tim, and Peter Schouten. 2001. ***A Gap in Nature: Discovering the World's Extinct Animals.*** New York: Atlantic

Monthly Press. 184pp. Well-written and thought-provoking look at some of the mammals, birds and reptile species definitely (or almost definitely) lost to the world since 1500. Beautifully illustrated with paintings by Shouten.

Gaal, Arlene. 2001. ***In Search of Ogopogo***. Blaine, WA: Hancock House. 208pp. Gaal's second book on the reported creature of Lake Okanagan includes a detailed chronology and many photographs. The author argues the presence of unknown animals "can no longer be disputed."

Harrison, Paul. 2001. ***Sea Serpents and Lake Monsters of the British Isles***. London: Robert Hale. 253pp. Harrison, president of the Loch Ness Monster Society, has written a readable collection of the more interesting accounts of the titular creatures, arranged geographically. A worthwhile contribution, although maps and footnotes would have strengthened it considerably.

Eberhart, George. 2002. ***Mysterious Creatures: A Guide to Cryptozoology***. Santa Barbara, CA: ABC-CLIO. 722pp. This massive two-volume encyclopedia attempts to catalog every cryptid worth mentioning, from purely mythical animals to those recently confirmed and described. References are given in all cases, and several supplementary essays cover important aspects of cryptozoology. The $185.00 price will limit its availability, but it's a most valuable resource for any zoologist or cryptozoologist. If you can afford it, get it.

Montgomery, Sy. 2002. ***Search for the Golden Moon Bear***. New York: Simon and Schuster. 336pp. Naturalist/writer Montgomery recounts her adventurous travels with biologist Gary Galbreath through Southeast Asia in search of a rare bruin with a stunning golden coat. A beautifully written travelogue as well as a book full of zoological information and discovery.

Paddle, Robert. 2002. ***The Last Tasmanian Tiger***. Cambridge, U.K.: Cambridge University Press. 284pp. A comprehensive account of the thylacine, especially the circumstances surrounding its head-on collision with human expansion. In addition to the factual record of the animal's deliberate extermination, Paddle offers his own theories on why the scientific world essentially ignored the thylacine until it was too late.

Reeves, Randall, *et al.* 2002. ***National Audubon Society Guide to Marine Mammals of the World***. New York: Knopf. 528pp. Thorough, sumptuously illustrated work covering all the marine mammals. Includes interesting recent developments on classification of the cetaceans. The lack of a bibliography is the only shortcoming.

Shuker, Karl. 2002. ***The New Zoo***. Thirsk (U.K.): House of Stratus. 304pp. An updated and expanded revision of Shuker's 1993 *The Lost Ark*, this book not only brings the story of new and rediscovered animals up to date, but adds much cryptozoological material and a large collection of illustrations. Very hard to find, but worth the effort.

Walker, John Frederick. 2002. ***A Certain Curve of Horn: The Hundred-Year Quest for the Giant Sable Antelope of Angola***. Boston: Atlantic Monthly Press. 320pp. Walker, a journalist, explores conservation, politics, hunting ethics, and other topics in the course of telling the story of the giant sable antelope. This distinctive subspecies of the sable antelope, instantly identifiable by its curving five-foot horns, was thought possibly extinct before its rediscovery in 2002.

Weidensaul, Scott. 2002. ***The Ghost with Trembling Wings***. New York: North Point Press. 341pp. Weidensaul explores and participates in searches for "missing" species, from birds to thylacines. He sometimes gives up hope for a species, as in the thylacine's case, although always with reluctance. A very valuable book packed with information and insight.

Coleman, Loren. 2003. ***Bigfoot! The True Story of Apes in***

America. New York: Pocket Books. 288pp. Coleman's book is a cogent (although still not, in my opinion, irrefutable) argument for the reality of an unclassified ape, as well as a good portrait of the often-fractious world of sasquatch hunters. The author can be a bit too enthusiastic, but good writing and thorough research still put this near the top of recent books on the subject.

Coleman, Loren, and Patrick Huyghe. 2003. ***The Field Guide to Lake Monsters, Sea Serpents, and Other Mystery Denizens of the Deep.*** New York: Penguin. 358pp. Similar to the same authors' book on mystery primates, this volume offers a classification system for all kinds of alleged denizens of seas, lakes, and rivers. Such beasts are so varied, though, that fourteen categories are needed, and even this approach requires grouping some unrelated reports together. I wish the authors had screened cases more skeptically, although they don't just swallow everything (the Surgeon's Photograph of Nessie, for example, is dismissed as a seal or otter). The sighting maps are a valuable addition. Bottom line: not perfect, but definitely useful.

Shuker, Karl P. N. 2003. ***The Beasts That Hide From Man***. New York: Paraview. 323pp. Eclectic collection of articles, essays, and notes on lesser-known cryptids, from small unclassified birds to the "death worm" of the Mongolian deserts. There is plenty of interesting reading, although the book would have benefited from better organization.

Arment, Chad. 2004. ***Cryptozoology: Science and Speculation***. Landisville, PA: Coachwhip. 393 pp. The first half of this book is a good exploration of the sources, methods, and scientific status of cryptozoology. The second is an exploration of cryptid reports, many little-known but intriguing (like the long-tailed wildcat that used to be reported in Pennsylvania). Definitely a worthwhile addition to the cryptozoological literature.

Coghlan, Ron. 2004. ***A Dictionary of Cryptozoology***. Bangor, Northern Ireland: Xiphis. 273pp. Hundreds of short entries, seemingly including every cryptozoological or mythical animal from around the world. A handy reference, especially for the more obscure reports.

Daegling, David. 2004. ***Bigfoot Exposed: An Anthropologist Examines America's Enduring Legend***. Walnut Creek, CA: AltaMira Press. 256pp. Daegling's most interesting argument is that the Patterson film is a hoax, but a sophisticated one that does not fit the story in Greg Long's book.

Ellis, Richard. 2004. ***No Turning Back: The Life and Death of Animal Species***. New York: Harper Collins. 428pp. Ellis' book on animal extinctions is readable, thought-provoking, and thoroughly researched, with some interesting tidbits for the cryptozoologist. It's quite a challenge to cover the whole topic of extinction, and some things are necessarily left out, but Ellis traces the modern cascade of extinctions and also finds space for a discussion of "back-breeding" efforts like those involving the Heck brothers' recreation of the aurochs. Ellis held out no hope for the ivory-bill and the Tasmanian tiger. He explains what has been going on in the world of cetaceans, where new species pop up while at least three plunge toward the extinction horizon. He discusses some other recent discoveries and rediscoveries as well. Ellis includes 41 pages of references and adds some footnotes throughout the book. The bottom line is that naturalists, nature writers, cryptozoologists, and zoologists will all want to read this one.

Hoose, Philip. 2004. ***The Race to Save the Lord God Bird***. New York: Farrar, Straus and Giroux. 208pp. Not as detailed as Jackson's book (below), but a touching, heavily illustrated look at the ivory-bill through its apparent demise in the U.S. and its last appearance in Cuba.

Jackson, Jerome. 2004. ***In Search of the Ivory-Billed Woodpecker***. Washington, D.C.: Smithsonian. 256pp. Second of two new books on the life, death, and possible survival of the magnificent ivory-bill, America's largest woodpecker, by an ornithologist who participated in the modern searches that turned up only hints and riddles until after this book was published.

Long, Greg, and Karl Korff. 2004. ***The Making of Bigfoot***. New York: Prometheus. 475pp. Long has put an enormous amount of research into this "unmasking" of the Patterson film, but in the end is not very convincing. He does a good job of pointing out problems with the accounts of Patterson and Gimlin, and shows convincingly that the events were not well investigated by cryptozoologists who were too quick to embrace the film. There is no convincing explanation, though, of the contradictory stories of the man who claims to have made the suit and the man who says he was in it. Cryptozoologists have pointed out errors in dates and locations that seem very odd for a book of such depth. In the end, sasquatch escapes, battered but not yet dead.

Murphy, Christopher L. (with John Green and Thomas Steenburg). 2004. ***Meet The Sasquatch***. Blaine, WA: Hancock House. 239 pp. This is a unique book featuring extremely good enlargements from the Patterson film and a lot of technical analysis of footprints, biomechanics, etc. This is essentially a brief for the prosecution: critics of sasquatch are summarily dismissed. Even for the most hardened skeptic, though, this book provides a lot to think about.

Owen, David. 2004. ***Tasmanian Tiger: The Tragic Tale of How the World Lost Its Most Mysterious Predator***. Baltimore: Johns Hopkins. 228pp. Owen presents a comprehensive account of what we know the thylacine's biology, behavior, and so on, weaving in the tale of its unfortunate interaction with human beings. He discusses the animal's two extinctions: the first on the Australian mainland, the second on Tasmania. Owen includes a discussion of recent sightings and the idea of bringing back the tiger via cloning.

Snyder, Noel. 2004. ***The Carolina Parakeet: Glimpses of a Vanished Bird***. Cambridge, MA: Princeton University Press. 176pp.

The first detailed account of the only native parrot of the United States. Snyder challenges the conventional wisdom that the bird was wiped out by hunting (disease may have played a bigger role) and presents sighting reports and other evidence indicating the species lingered for decades beyond its official 1914 extinction date.

Mittelbach, Margaret, and Michael Crewdson. 2005. ***Carnivorous Nights: On the Trail of the Tasmanian Tiger.*** New York: Villard Press. 336pp. A funny, sad, and definitely unique travelogue recounting journeys to the historic and rumored haunts of the thylacine. The authors don't provide the final word on the tiger's existence, but they introduce us to its world in a vivid fashion.

Newton, Michael. 2005. ***Encyclopedia of Cryptozoology: A Global Guide to Hidden Animals and Their Pursuers***. McFarland & Company. 576pp. Newton's 2,774 entries include some overlap with Eberhart's *Mysterious Creatures*, though this book is stronger on profiling groups and individuals. Like Eberhart's work, this is a good reference with an unfortunately steep price ($95.00).

Ellis, Richard. 2006. ***Singing Whales, Flying Squid, and Swimming Cucumbers: The Discovery of Marine Life***. New York: Lyons. 336pp. Ellis' book was not available in time for review, but his exploration of the title topic is sure to be worth reading.

PERIODICALS

This section presents a small sample of the most interesting and valuable articles published over the last two decades. Some species descriptions and other technical articles are included in this section, but I've focused mainly on publications likely to be available to the general reader. These are grouped by year of publication and then listed alphabetically by author. Many of the more technical articles are cited in the References section in connection with the essays they were used to substantiate.

PERIODICALS 1983-1992

Begley, Sharon, with John Carey. 1983. "In Search of Mythic Beasts," ***Newsweek***, April 11, p.74. Article on the ISC and its quests.

Simpson, George Gaylord. 1984. "Mammals and Cryptozoology," ***Proceedings of the American Philosophical Society*** 128, p.1. The late (1902-1984) paleontologist had no patience with cryptozoology, which he did not consider a science at all. In this article, Simpson predicted there were no significant mammal discoveries to be made in the future—a mistake of enormous proportions. This article should still be read for its background on mammal discoveries and to understand the problems some scientists have with the term "cryptozoology."

Diamond, Jared. 1985. "How many unknown species are yet to be discovered?" ***Nature*** (315), June 13, p.538. Review of discoveries from 1900 to 1983, with emphasis on mammals.

Diamond, Jared. 1985. "In Quest of the Wild and Weird," ***Discover***, March, p.35. Well-written, generally skeptical review of cryptozoology.

Wolkomir, Richard. 1986. "Tracking Down Monsters," ***International Wildlife***, March-April, p.24. More sympathetic examination of the field of cryptozoology.

Belfield, Dominic. 1987. "Of dinosaurs, bunyips, and yeti," ***New Scientist***, August 27, p.59. Article encouraging scientists to keep an open mind about new animals, even the "monsters."

Carr, Archie. 1987. "The Age of Discovery Revisited," ***Animal***

Kingdom, January/February, p.35. Report on recent animal discoveries around the world.

James, Jamie. 1988. "Bigfoot or Bust," ***Discover***, March, p.44. Generally positive look at cryptozoology (the unicorn in the lead illustration was presumably meant to be humorous).

Bolgiano, Chris. 1991. "Concepts of cougar," ***Wilderness***, Summer. The lore of cougars, including information on the Eastern cougar.

May, Robert M. 1992. "How Many Species Inhabit the Earth?" ***Scientific American***, October, p.42. Thorough review of the title question.

PERIODICALS 1993

Anonymous. 1993. "Bizarre Bovid Identified in Vietnam," ***Science News Letter***, June 19, p.397. Well-written recounting of the Vu Quang ox story.

Dennis, Jerry, and Glenn Wolff. 1993. "Dinosaur Hunting in Our Time," ***Wildlife Conservation***, July-August, p.72. Balanced examination of cryptozoology in general.

Huyghe, Patrick. 1993. "New-Species Fever," ***Audubon***, March-April, p.88. Survey of recent discoveries and ongoing searches.

McCarthy, Paul 1993. "Cryptozoological Quests," and "Cryptozoologists: an Endangered Species." ***The Scientist*** (7:1). In a pair of articles, McCarthy described the travails of scientists trying to find support for cryptozoological research. A follow-up letter (in ***The Scientist*** 7:8) describes how evolutionary biologists do much the same thing cryptozoologists do—speculate about and search for unproven forms of life. As long as they stick to extinct forms, though, they avoid the ridicule accompanying cryptozoology.

Monaghan, Peter. 1993. "Cryptozoologists Defy Other Scientists' Skepticism to Stalk Beasts Found in Legend, Art, and History," ***The Chronicle of Higher Education***, February 10, p.A7. Discusses cryptozoology's reception in academia.

Rogers, Michael. 1993. "Little Birds Lost," ***National Wildlife***, October/November, p.20. Review of the uncertain status of

Hawaii's native honeycreepers, including recently-discovered, rediscovered, and possibly extinct species.

PERIODICALS 1994

Boston, Robert. 1994. "Is Bigfoot an Endangered Species?" ***Skeptical Inquirer***, Fall, p.528. Scathing and one-sided review of Dr. Grover Krantz's pro-Sasquatch book *Big Footprints*.

Conover, Adele. "Did Columbus Sup on the World's Last Giant Sloth?" ***Smithsonian***, October, p.20. History of South America's ground sloths, with discussion of the mapinguari.

Dennett, Michael R. 1994. "Bigfoot Evidence: Are These Tracks Real?" ***Skeptical Inquirer***, Fall, p.498. Alleged sasquatch footprint casts from Indiana are exposed as a hoax.

George, Carol Ann. 1994. "Featured Feline: Iriomote Cat," ***Cat Tales***, April, p.5. The status of this critically endangered feline (*Felis iriomotensis*), described only in 1967.

Greenwell, J. Richard. 1994. "Prehistoric Fishing," ***BBC Wildlife***, March, p.33. Recent events in the study of the coelacanth and its possible range beyond the Comoros Islands.

Karesh, William. 1994. "Update from Vietnam," ***Wildlife Conservation***, November/December, p.16. Status of another rediscovery, the Tonkin snub-nosed monkey (*Pygathrix avunculus*).

Lutz, Richard A., and Rachel Haymon. 1994. "Rebirth of a Deep-sea Vent," ***National Geographic***, November, p.115. The submersible *Alvin* revisits a vent devastated by a volcanic eruption and finds its fauna recovering quickly. Includes photo of a new species of octopus.

Morland, Hilary. 1994. "Eagle photo a First," ***Wildlife Conservation***, November/December, p.16. First photograph of a Madagascar serpent eagle, rediscovered after presumed extinction.

Nobbe, George. 1994. "Forget the Fins—This Caddy's Got a Flipper," ***OMNI***, August, p.26. Roundup of news and research concerning "Caddy," the large unclassified sea animal believed to exist off the west coast of British Columbia.

Wise, Jeff. 1994. "If a Sao La is Seen in the Forest, But Not By

Scientists, Does it Exist?" ***New York Times Magazine***, p.38. In-depth article on John MacKinnon's search in Vu Quang.

PERIODICALS 1995

Anonymous. 1995. "Smoking Worms," ***Discover***, July, p.24. Update on the tube worms which cluster around the "black smokers" on the seafloor.

Anonymous. 1995. "Lost World of New Creatures Found in a Romanian Cave," ***National Geographic***, October. Surveys the bizarre inhabitants of recently-discovered Movile Cave, where the sunless, low-oxygen environment supports dozens of unique species of invertebrates.

Bearder, Simon K. 1995. "Calls of the Wild," ***Natural History***, August, p.48. Report on the identification of no fewer than five new species of the small African primates called galagos, or bush babies, based on recording and analysis of their calls.

Hendrix, Steve. 1995. "Quest for the Kouprey," ***International Wildlife***, September/October, p.20. Status of the endangered kouprey, the largest land animal discovered in the 20th century.

Klos, Ursula and Heinz-Geog. 1995. "Are There White Elephants?" ***Elephant Managers Association Newsletter***, October, p.61. The white elephant in myth and reality, with photos.

Magsalay, Perla, et al. 1995. "Extinction and Conservation on Cebu," ***Nature***, January 26, p.294. Status of rediscovered but apparently extinction-bound birds on the island of Cebu.

Popular Science (various authors). Special issue, "Our Ocean Planet," May 1995. Includes articles on fauna of the seafloor vents and on the elusive giant squid.

Royte, Elizabeth. 1995. "Hawaii's Vanishing Species," ***National Geographic***, September, p.2. Well-illustrated look at Hawaii's rare endemics, some recently discovered or presumed extinct.

Shuker, Karl P. N. 1995. "The Florida Globster—Verilly, a mystery," ***Wild About Animals***, September, p.9. Good recap of the *Octopus giganteus* controversy.

Underwood, Anne. 1995. "Where Have All the Malas Gone?" ***International Wildlife***, March/April, p.14. Australia's endangered marsupials, including the bridled nailtail wallaby, thought extinct until 1973.

Whitehead, Hal. 1995. "The Realm of the Elusive Sperm Whale," ***National Geographic***, November, p.57. This article includes photographs of a white sperm whale calf and one of a whale's tail from an angle that makes it look like the head and neck of an unknown animal.

PERIODICALS 1996

Conniff, Richard. 1996. "Clyde Roper can't wait to be attacked by the giant squid," ***Smithsonian***, May, p.126. Profiles Dr. Roper's quest to find Architeuthis. Conniff notes that Roper has already "identified one new family, two genera and about twenty species."

Etter, Matt, and Nancy Ruggeri. 1996. "First Adult Saola Sighted," ***Wildlife Conservation***, March/April, p.9. First report of an adult Vu Quang ox, or saola, captured alive.

Schaller, George B., and Elisabeth S. Vrba. 1996. "Description of the Giant Muntjac (*Megamuntiacus vuquangensis*) in Laos," ***Journal of Mammology***, 77(3), p.675. Formal description of one of the new land mammals from SE Asia.

Stolzenburg, William. 1996. "The Lost World, Part II," ***Nature Conservancy***, May/June, p.7. Quick review of the startling discoveries in the animal kingdom over the last few years.

PERIODICALS 1997

Discover (various authors). 1997. January. This issue had three articles of interest. "Shell Game" concerns the discovery of the Gulf snapping turtle. "A Marmoset With Appendages" describes the new primate *Callithrix saterei*. "Where the Running Rodents Play" discusses the Panay cloudrunner (*Crateromys heaneyi*), another new mammal described in 1996.

Eliot, Jon. 1997. "In the Dwindling Philippine Forest, There's Still Room for Discovery," ***National Geographic***, July. Article with photos of two new species: the Panay cloudrunner, a squirrel-sized mammal, and Lina's sunbird, just identified from specimens taken in 1965.

Fortean Times (various authors). 1997. July. I'm reluctant to cite publications that mix cryptozoology with unrelated subjects like psychic phenomena, but this issue offers four articles on cryptozoology, all of them worth reading. "Hell's Teeth," by Jeremy Wade, concerns a bizarre animal Wade photographed in the Amazon: a dolphin with a weird saw-toothed ridge down its back. Dr. Karl Shuker's column, "Lost Ark," tells the story of the rediscovery of Borneo's bay cat after 60 years. Loren Coleman writes on claims of giant anacondas, and there's an update on the strange chimp "Oliver." The July issue follows up with Karl Shuker's Top Twenty animal discoveries of recent years, a recommended Top Ten cryptozoological books, and a photograph of a koolookamba, an alleged chimp-gorilla cross that does display curious features.

Horvitz, Leslie. 1997. "Cryptozoologists Try to Separate Strange Fact From Science Fiction," ***Insight***, January 27, p.44. Sympathetic and well-written overview of cryptozoology.

Line, Les. 1997. "Phantom of the Plains," ***Wildlife Conservation***, August, p.20. Update on the black-footed ferret (*Mustela nigripes*), which was twice rediscovered after presumed extinction.

Mittermeir, Russell. 1997. "Homegrown Varieties," ***Newsweek***, August 11, p. 20. Dr. Mittermeir writes that, in our excitement over possible extraterrestrial life, we still need to catalogue many species living on Earth. He cites new mammals from Brazil as an example.

MacPhee, Ross, and Clare Flemming. 1997. "Brown-eyed, milk-giving…and extinct," ***Natural History***, April, p.84. Sobering study of mammalian extinctions over the last 500 years.

Mlot, C. 1997. "Newfound worm's world under the sea," ***Science News***, August 9, p.86. Short article on the bizarre worms found living on frozen methane hydrate, with striking photographs.

Newman, Cathy. 1997. "Nature's Masterwork: Cats," ***National Geographic***, June, p.54. Reclassification of the cat tribe. Newman believes the puma belongs in its own genus (Puma), as does the most recent new cat species, the Iriomote cat (*Mayailurus iriomotensis*).

Quammen, David. 1997. "You Looking for Me?" ***Sports Illustrated***, February 3, p.66. Quammen recounts an Amazon expedition to confirm a new species of monkey.

Sayers, Kenneth. 1997. "Is It Over Yeti?" ***Skeptical Inquirer***, March/April, p.51. Thorough, fair-minded review of Daniel Taylor-Ide's 1995 book *Something Hidden Behind the Ranges*.

Shuker, Karl. 1997. "A Surfeit of Civets?" ***Fortean Times***, September, p.17. Dr. Shuker reports on evidence for two possible new species of civets. In the same issue, Mike Dash has a superbly researched, balanced article surveying "lake monsters" from around the world.

Tangley, Laura. 1997. "New Mammals in Town," ***U.S. News and World Report***, June 9, p.59. Well-written introduction to the recent explosion in mammal discoveries.

PERIODICALS 1998

Allen, Thomas. 1998. "Deep Mysteries of Kaikoura Canyon," ***National Geographic***, June, p.106. Chronicle of the efforts to study giant squid off New Zealand, with underwater photos.

Burney, David, and Ramilisonina. 1998. "The Kilopilopitsofy, Kidoky, and Bokyboky: Accounts of Strange Animals from Belo-sur-mer, Madagascar, and the Megafaunal 'Extinction Window,'" ***American Anthropologist*** 100:4 (December), p.957. Fascinating study of Madagascar folklore indicating that animals known only as fossils lived into historical times or may still be living.

Colla, Phillip, and Harrison Stubbs. 1998. "Guadeloupe's Regal Fur Seal." ***Ocean Realm***, Summer, p.60. Saga of a pinniped that was twice believed hunted to extinction.

Line, Les. 1998. "Is This the World's Rarest Bird?" ***National Wildlife***, December/January, p.46. Plight of the nearly extinct po'ouli, the Hawaiian honeycreeper discovered only in 1973.

Schaller, George. 1998. "On the Trail of New Species," ***International Wildlife***, July/August, p.37. Schaller's adventures and discoveries in Southeast Asia.

Tenneson, Michael. 1998. "Expedition to the Clouds," ***International Wildlife***, March/April, p.32. Chronicle of a biological expedition to the cloud forests of Peru. Discoveries so far include new birds, twelve new reptiles, and a large new rodent.

Teresi, Dick. 1998. "Monster of the Tub." ***Discover***, April, p.86. Examination of the Lake Champlain monster phenomenon. Goes into detail about the seiche wave phenomenon (which works like water sloshing in a bathtub, hence the article's title).

PERIODICALS 1999

Boroughs, Don. 1999. "New Life for a Vanished Zebra?" ***International Wildlife***, March/April, p.46. Recounting of Reinhold Rau's determined efforts to breed the quagga back from oblivion.

Daegling, David. 1999. "Bigfoot's Screen Test," ***Skeptical Inquirer***, May, p.20. Thorough explanation of the problems in attempting to authenticate the 1967 Patterson sasquatch film.

Holloway, Marguerite. 1999. "Beasts in the Mist," ***Discover***, September, p.58. Story of David Oren's search for Amazonia's mapinguari, which Oren thinks is a surviving ground sloth.

Graham, Rex. 1999. "In Search of Ecuador's Newest Bird," ***Birder's World***, June, p.44. The author accompanies Robert Ridgely on a search for the newly discovered jocotoco.

Peck, Robert McCracken. 1999. "Home Again!" ***International Wildlife***, September/October, p.36. The saga of Przewalski's horse and its rescue from near-extinction.

Raloff, Janet. 1999. "Rarest of the Rare," ***Science News***, September 4, p. 153. Rediscovery and conservation of the mainland population of the Javan rhinoceros.

Stewart, Doug. 1999. "Sponges Get Respect," ***International Wildlife***, July/August, p. 26. Article on the phylum Porifera, including the recently-discovered carnivorous sponge.

Zorpette, Glenn. 1999. "Chasing the Ghost Bat," ***Scientific American***, June, p.82. An expedition studies Belizean bats and records calls of possible unknown species.

PERIODICALS 2000

Alexander, Charles. 2000. "Death Row," ***TIME***, January 17, p.76. Photographic essay on the twenty-five most endangered primates—eight of them discovered or rediscovered in the last three decades.

Bindernagel, John. 2000. "Sasquatches in our woods," ***Beautiful British Columbia***, Summer, p.28. Bindernagel, a biologist, argues that sasquatch should be taken seriously.

Erdmann, Mark. 2000. "New Home for 'Old Fourlegs,'" ***California Wild***, Spring, p.8. Erdmann's first-hand account of the discovery of the second coelacanth species.

Freeman, Mark. 2000. "The Man Who Saw Bigfoot," ***Wildlife Journal***, Winter, p.68. Overall, a very good article on the Matthew Johnson sasquatch sighting, although the author inexplicably wastes some space on a sasquatch-hunter propounding a "supernatural entity" theory.

Friederici, Peter. 2000. "Colorless in a World of Color," ***National Wildlife***, p.44. Intriguing story about albinistic and leucistic animals, with photographs of species from birds to moose.

Holloway, Marguerite. 2000. "Save the Muntjacs," ***Scientific Ameÿrican***, December 2000. The discoveries and conservation efforts of Alan Rabinowitz in Southeast Asia and elsewhere.

Lutz, Richard. 2000. "Deep Sea Vents," ***National Geographic***, October, p.116. Illustrated update on the vents which have produced "a new species every week and a half" for twenty-one years.

McClintock, Jack. 2000. "20 Species We May Lose in the Next 20 Years," ***Discover***, October, p.62. The world's most endangered species, including the saola.

Messner, Reinhold. 2000. "Encounter With a Yeti," ***National Geographic Adventure***, June, p.130. Messner recounts his sighting of a yeti. Includes an interview with explorer Johan Reinhard, who considers the yeti a myth but mentions his own sighting of unexplained tracks.

Naish, Darren. 2000. "Where Be Monsters?" ***Fortean Times***, March, p.40. Excellent critique of the many overly enthusiastic cryptozoological theories about sea serpents.

Simpson, Sara. 2000. "Looking for Life Below the Bottom," ***Scientific American***, June, p.94. Expedition to collect samples of the countless microscopic life forms hiding beneath the seafloor.

Warshall, Peter, *et al.* 2000. ***Whole Earth*** magazine, Fall. Special issue on the All Species Inventory project, a grandiose endeavor to catalog the world's entire biota in twenty-five years. Includes articles on new marmosets (by David Quammen), new birds (Don Stap), and basic taxonomy (Terry Erwin), along with many other valuable pieces.

PERIODICALS 2001

Boroughs, Don. 2001. "Not as Dead as a Dodo," ***International Wildlife***, November/December. p.44. The resuscitation of three nearly-extinct birds of Mauritius Island, former home of the dodo.

Bruemmer, Fred. 2001. "Comeback on a Castaway's Island," ***International Wildlife***, March/April, p.38. Current state of the Juan Fernandez fur seal, a species once thought extinct.

Elton, Catherine. 2001. "Evangelist for Nature." ***International Wildlife***, May/June, p.30. Adventures of conservationist Jose Alvarez, including his discovery of several new birds.

Flannery, Tim. 2001. "A Lost Menagerie," ***Natural History***, November, p.66. Excerpt from Flannery's book *A Gap in Nature*, portraying several lost or mystery species.

Frasier, Caroline. 2001. "The Ballad of Lonesome George," ***Outside***, January. The evolution and conservation of species in the Galapagos islands. (Lonesome George is the very last Pinta Island giant tortoise, standing a kind of deathwatch until his kind goes extinct.)

Guynup, Sharon. 2001. "Dr. Rabinowitz, I Presume," ***Wildlife Conservation***, March/April, p.36. The travels of Alan Rabinowitz, zoologist, conservationist, and discoverer of the leaf muntjac and other new species.

Lundberg, John. 2001. "Freshwater Riches of the Amazon," ***Natural History***, September, p.36. The amazing diversity of Amazonian fish life, including information on some of the startling new species the author's expeditions discovered.

Moffett, Mark. 2001. "Travels with Charlie," ***International Wildlife***, March/April, p.44. An entomologist explores Venezuela's tepuis for new species.

Nickell, Joe. 2001. "Tracking the Swamp Monsters," ***Skeptical Inquirer***, July/August. Looking for sasquatch-type creatures in southern American swamps.

Perkins, Sid. 2001. "The Latest Pisces of an Evolutionary Puzzle," ***Science News***, May 5, p.282. The news of new coelacanth discoveries and the puzzles concerning the fish's range.

Tangley, Laura. 2001. "Mysteries of the Twilight Zone," ***National Wildlife***, October/November, p.52. The emerging picture of animal diversity in the abyssal depths.

Turnbull, March. 2001. "Back from the Dead?" ***Africa***, April, p.30. The latest on the controversy over the quagga and the efforts to recreate it through selective breeding of zebras.

Vecchione, Michael. 2001. "Worldwide observations of remarkable deep-sea squid," ***Science*** (294), p.2505. Introduces a startling new animal—the huge-finned, 20-foot mystery squid.

PERIODICALS 2002

Adis, Joachim, et al. "Gladiators: A New Order of Insect," ***Scientific American***, November, p.60. Story of the discovery of the first new insect order in many decades, the large and spectacular "gladiators" of Africa.

Boyd, Lee. 2002. "Reborn Free," ***Natural History***, July/August, p.56. The reestablishment in the wild of Przewalski's horse, brought back from near-extinction.

Busch, Robert. 2002. "Squidzilla," ***Wildlife Conservation***, February, p.30. The latest discoveries about one of cryptozoology's favorite creatures, the giant squid Architeuthis.

Eastabrook, Barry. 2002. "Staying Alive," ***Wildlife Conservation***, May/June, p.36. Examination of the options, including cloning, for saving or restoring species.

Jackson, Jerome. 2002. "The Truth is Out There," ***Birder's***

World, June, p.40. Frustration and hope in the search for the ivory-billed woodpecker.

Klass, Klaus, et al. 2002. "Mantophasmatodea: a new insect order with extant members in the Afrotropics," ***Science***, May 24, p.1456. This article names the first new order of living insects since 1914.

Milius, Susan. 2002. "Are They Really Extinct?" ***Science News***, March 16. The search for plants and animals presumed extinct, with some recent success stories.

Pitman, Robert. "2002. "Alive and Whale," ***Natural History***, September, p.32. Confirmation of the mysterious beaked whale, *Mesoplodon pacificus*.

Schilthuizen, Menno. 2002. "Caution: Species Crossing," ***Natural History***, September, p.62. Examples of the effects and difficulties created for taxonomists by hybridization in the wild.

Weidensaul, Scott. 2002. "In Search of the Phantom Tanager," ***Outside***, June. A search of Brazil's Mato Grosso region for the cone-billed tanager, collected in 1938 and never seen again.

Wildlife Conservation. 2002. "Carnivore Rediscovered in Tanzania," September/October, p.8. Article with photograph of Long's servaline genet, missing since 1932.

PERIODICALS 2003

Duffy, J. Emmett. 2003. "Underwater Urbanites," ***Natural History***, December 2003-January 2004, p.40. Review of the eusocial shrimps of the genus Synalpheus, many species of which have only recently been described.

Nickell, Joe. 2003. "Legend of the Lake Champlain Monster," ***Skeptical Inquirer***, July/August, p.18. Argument for "non-monster" explanations for the creature of Lake Champlain. The accompanying article by Benjamin Radford, "The Measure of A Monster," casts doubt on the interpretation of the Mansi photograph.

Pitman, Robert. 2003. "Good Whale Hunting," ***Natural History***,

December 2003-January 2004, p.24. Account of Pitman's observations of Antarctic killer whales, which indicate one or even two undescribed species may be present.

Sieveking, Paul. 2003. "Big Cats in Britain," ***Fortean Times***, March, p.28. Thorough examination of the phenomenon of alien or unknown big cats reported all over the British Isles.

Sterling, Eleanor, et al. 2003. "Vietnam's Secret Life," ***Natural History***, March, p.50. Complete story of recent developments in Vietnamese biodiversity, including discoveries made since 1990.

Vangelova, Luba. 2003. "True or False? Extinction is Forever," ***Smithsonian***, June, p.22. The challenges and implications of bringing back the thylacine via cloning.

PERIODICALS 2004

Coleman, Loren. 2004. "Costume Drama," ***Fortean Times***, August, p.58. In-depth (and not complimentary) review of Greg Long's book "exposing" Bigfoot.

Hayden, Thomas, et al. 2004. "Blue Planet," ***U.S. News & World Report***, August 23, p.46. Series of articles including text and photos of new and strange deep-water creatures.

Gran, David. 2004. "The Squid Hunter," ***The New Yorker***, May 24, p57. Following the quest of marine biologist Steve O'Shea for knowledge of the giant squid.

Korff, Karl, and Michael Kocis. 2004. "Exposing Roger Patterson's 1967 Bigfoot Film Hoax," ***Skeptical Inquirer***, July/August, p.35. Report on Long's claimed exposure of the Patterson film.

Loxton, Daniel. 2004. "Big Foot Con," ***Skeptic***, Vol. 11, No. 1, p.82. Review of Greg Long's book on Bigfoot. The same issue of Skeptic contains an extensive ***Junior Skeptic*** section, also by Loxton, ridiculing the Loch Ness monster.

Mairson, Alan. 2004. "Into the Wild," ***National Geographic***, October, p.106. Scientists collect new reptile species in the deserts of Iran.

Shuker, Karl. 2004. "The not so silly Bili Ape.." ***Fortean Times***,

October. P.56. Review of the puzzling saga of the chimplike apes of the Bili region.

Snifka, Lynne. 2004. "Monstrous Mysteries," ***Alaska***, October. The mystery beasts of Alaska, from the Lake Iliamna monster to the Hairy Man.

Young, Emma. 2004. "The beast with no name," ***New Scientist***, October 9, p.33. Thorough examination of the Bili ape mystery.

Harrison, Bobby. 2005. "Phantom of the Bayou," ***Natural History***, September, p.18. First-hand account of the ivory-bill's rediscovery.

Fitzpatrick, John, et al. 2005. "Ivory-billed Woodpecker (*Cempephilus principalis*) Persists in Continental North America," ***Science*** 3, June, p.1460. The paper that rocked the zooloogical world.

Kubodera, Tsunemi, and Kyoichi Mori. 2005. "First-ever observations of a live giant squid in the wild," *Proceedings of the Royal Society*. Published online (http://www.journals.royalsoc.ac.uk), September 27.

Leiberman, D.E. 2005. "Further fossil finds from Flores," ***Nature*** 437, October 13, p.957.

Morwood, M.J., et al. 2005. "Further evidence for small-bodied hominins from the Late Pleistocene of Flores, Indonesia," ***Nature*** 437, October 13, p.1012.

Wilford, John. 2005. "A Big Debate on Little People: Ancient Species or Modern Dwarfs?" ***New York Times***, October 12.

INTERNET SITES

The Internet holds an uncountable and ever-changing number of sites of interest, and no essay like this one can present more than a fraction of one percent. The Net is home to both cryptozoology sites and "mainstream" zoology sources. All Internet sites were checked before this book went to press to verify they were still functional. The Internet changes so frequently, though, there's no way to guarantee all these sites will still be good at the time this book is read.

The line between zoological and cryptozoological sites is blurry, since museums and other staid institutions host data on such items of cryptozoological interest as the giant squid and the Vu Quang ox.

Speaking of Vu Quang, there's plenty of information on the discoveries here on the Net. Vern Weitzel of Australia has posted information and photos on the Vu Quang ox, the giant muntjac, and Vo Quy's pheasant at http://coombs.anu.edu.au/~vern/species.html. A data sheet on the odd ox is available from the World Conservation Monitoring Centre at: http://www.wcmc.org.uk/species/data/species_sheets/vuquang.htm.

To give a few more examples, a good collection of resources on rare and endangered mammals can be found at http://www.animal-info.org. The All Species Foundation, which supports a project intended to classify all the plants, animals, fungi, and microbes of the Earth, is at www.all-species.org. A useful index, with free abstracts, is available for the Journal of Mammology at the American Society of Mammology page at http://www.mammalsociety.org/publications/, and a linked page offering free downloads of the full descriptions of 631 mammal species is at http://www.science.smith.edu/departments/Biology/VHAYSSEN/msi/.

Other sources for information on rare and endangered animals can be found at http://www.enn.com (the Environmental News Network), http://www.eelink.net (EE-Link), and the zoology page of ***Science Daily***: (http://www.sciencedaily.com/odp/Top/Science/Biology/Zoology).

The Animals and Nature page of ***National Geographic News*** (http://news.nationalgeographic.com/news/animals.html) is always valuable.

The home page for Conservation International (www.conservation.org) includes news along with information on CI's work and a variety of other resources. Numerous papers on the results of CI's RAP team investigations (discussed in Section I) are available online. The World Wide Fund for Nature, a.k.a. the World Wildlife Fund, also has a good site (www.panda.org). The International Union for the Conservation of Nature has its own page at www.iucn.org. The IUCN is famous for its "Red List" of threatened species. The 2004 list is posted at www.redlist.org. The IUCN depends for much of its information on rare and threatened birds on BirdLife international, which posts news and other information at www.birdlife.net.

The uniquely named ***WorldTwitch*** site at http://worldtwitch.virtualave.net collects news of new and rare species of birds. An outstanding source for primate data is the Primate Info Net site at http://www.primate.wisc.edu/pin/. The ***Cetacea*** site at http://www.cetacea.org has excellent descriptions of every species, including the mysterious Mesoplodon "Species A." Jerome Hamlin's site ***Coelacanth: The Fish Out of Time***, at http:www.dinofish.com, is dedicated to the conservation of the world's most famous fish.

There are certainly plenty of sites related to cryptozoology. Punching "cryptozoology" into the Google search engine alone brings up 1,030,000 matches (including a company called Sasquatch Music).

A useful cryptozoology Internet publication is the ***North American BioFortean Review***. Editors Chad Arment and Brad LaGrange seem a little less skeptical of eyewitness accounts than I'd like, but that's a judgment call. NABR is posted at Chad's site at www.strangeark.com and can be downloaded in pdf format. It's a good publication, with a solid roster of contributing authors. The site also features a wide range of cryptozoological data and even some .pdf files with the text of relevant books (only those old enough to be unprotected by copyright law, of course).

A long-running cryptozoological site is maintained by Canadian enthusiast Ben Roesch at: http://www.ncf.carleton.ca/~bz050/HomePage.cryptoz.html. His site includes links, articles, and other information, with a special emphasis on Ben's favorite subject, sharks. Also very thorough is Scott Norman's ***Cryptozoology Realms***, http://www.cryptozoologicalrealms.com. ***Cryptoozoology.com*** (at, of course, http://www.cryptozoology.com), run by William Duncan, Jim Harnock, and Cisco Serret, is always a favorite for keeping up with news from around the world. Other top sites are "Pib" Burns' ***Cryptozoology Resources*** (http://www.pibburns.com/cryptozo.htm) and Michel Raynal's ***Virtual Institute of Cryptozoology*** (in English and French) at http://perso.wanadoo.fr/cryptozoo/welcome.htm. The official International Society of Cryptozoology site at http://www.cryptozoologysociety.org, like the Society itself, still exists but seems moribund. The British Columbia Scientific Cryptozoology Club site, http://www.bcscc.ca.tt/, is excellent, as is Crypto mundo at http://www.cryptomundo.com.

One of the leading crypto-authors, Dr. Karl Shuker, is on the Web at http://members.aol.com/karlshuker. Not to be outdone, I have my own site at http://www.mattwriter.com.

Loren Coleman's extensive site, ***The Cryptozoologist***, http://www.lorencoleman.com/, is well worth a visit. Loren has created the first cryptozoology museum in Portland, Maine.

Charles Paxton has erected ***The Aquatic Monsters Site*** at www.sea-monster.org. It may not be the catchiest name, but Paxton is a well-credentialed scientist and his thoughts worth reading.

Those are a few of the general cryptozoology sites. There are many devoted to specific cryptids or types of cryptids. For example, ***The Legend of Nessie*** offers one of the best Loch Ness sites at http://www.nessie.co.uk. with many reports and illustrations. Adrian Shine's site at www.lochnessproject.org is well worth a visit. Dick Raynor's site at http://www.dickraynor.co.uk has a mass of data and a well-balanced viewpoint, including a lot of technical information on photography, conventional explanations, and other reference topics. At www.pbs.org/wgbh/nova/lochness/legend.html, you can read about and buy the video of the documentary, "The Beast of Loch Ness," the best TV program ever on this subject. The Smithsonian Institution has its own Loch Ness site, with a good bibliography, at http://www.si.edu/resource/faq/nmnh/lochness.htm. To see the Loch for yourself, go to the Loch Ness Live Webcam at http://www.lochness.co.uk/livecam/index.html. Finally, it's not an Internet resource, but I should mention that Andreas Trottman runs a useful ***Loch Ness Newsclipping Service*** (address: Andreas Trottman, Les Pretresses, 1586 Vallamand VD, Switzerland).

For information on the Dr. John Bindernagel's Sasquatch book, readers can check his Web site at http://www.bigfootbiologist.org. The Bigfoot Field Researchers Organization can be found at http://www.bfro.net. The BFRO Web site offers a huge database of sightings. Two other sasquatch sites worth visiting are the Gulf Coast Bigfoot Research Organization's site, www.gcbro.com, devoted to the "southern bigfoot," and the Washington State Sasquatch Search Group's site at : http://www.wsssg.net.

The ***Champ Quest*** site at http://www.champquest.com focuses on the beasts reported to prowl the depths of Lake Champlain. Of special note at this site is the collection of photographs and downloadable video clips.

For those curious about the Eastern cougar, a good starting point is the Eastern Cougar Foundation page at http://www.easterncougar.org.

As I said, these are just the sites to begin with. There are countless others, and together they offer a gigantic library of information (good and bad). Keep in mind anyone can post anything on the internet, and a good writer can make even totally unsubstantiated material sound convincing. The Internet is an invaluable tool, but use it with caution.

ORGANIZATIONS

Two types of organizations are listed here: cryptozoological ones and mainstream zoology/conservation groups.

The International Society for Cryptozoology (ISC) was founded to bring scientific rigor and respect to cryptozoology. Its President was the original cryptozoologist, the late Dr. Bernard Heuvelmans. With the 2005 passing of cofounder J. Richard Greenwell, the ISC appears all but gone. Nevertheless, the ***ISC Newsletter*** and the peer-reviewed journal ***Cryptozoology*** were first-rate contributions and should be sought out by any serious students of this topic. The ISC site at http://www.cryptozoologysociety.org still exists.

The British Columbia Scientific Cryptozoology Club (BCSCC) has filled part of the gap caused by the ISC's difficulties. Headed by President John Kirk and Dr. Paul LeBlond, it covers all types of cryptids in its publications but gives special emphasis to aquatic cryptids and sasquatch. The BCSCC's quarterly newsletter does give appropriate space to reporting prosaic explanations for cryptids and minimizes unfounded speculation.

Contact:
John Kirk, President
BCSCC, Suite 89,
6141 Willingdon Avenue,
Burnaby, British Columbia, CANADA V5H 2T9.
E-mail: bccryptoclub@yahoo.com
http://www.bcscc.ca.

Then there are the mainstream zoological organizations, which support much of the work that conserves rare species and finds new ones.

The Wildlife Conservation Society (WCS) is best-known as the parent organization of the Bronx Zoo. WCS conservation and research programs operate in over 50 countries. Individual membership includes the excellent magazine ***Wildlife Conservation***.

Contact:
Wildlife Conservation Society
2300 Southern Boulevard
Bronx, NY 10460-1099
Telephone: (718) 220-5111 weekdays

Email: membership@wcs.org
Website: http://www.wcs.org

The National Wildlife Federation is a leader in conservation and publishes two magazines, ***National Wildlife*** and ***International Wildlife***. Associate Membership includes a choice of magazines.

Contact:
National Wildlife Federation
11100 Wildlife Center Drive
Reston, VA 20190-5362
Telephone: (800) 822-9919
Website: http://www.nwf.org

The WWF, known also as the World Wildlife Fund or the World Wide Fund for Nature, is another international leader in supporting conservation and research. It includes five million supporters and twenty-seven national programs. Individual membership begins at $15.

Contact:
World Wildlife Fund
1250 Twenty-Fourth Street, N.W.
P.O. Box 97180
Washington, DC 20077-7180
Telephone: (800) CALL-WWF
Websites:
U.S. - http://www.worldwildlife.org
International - http://www.wwf.org or www.panda.org

Conservation International supports the RAP and AquaRAP teams described in Section I. CI is not a membership organization in the sense the others are, but invites people to become members by making a gift of any size to their "Campaign to Save the Hotspots."

Contact:
Conservation International
1919 M Street, NW Suite 600
Washington, DC 20036
Telephone: (202) 912-1000
Website: http://www.conservation.org

Flora and Fauna International (FFI) is a global conservation organization founded in 1903. Among many other projects, it supports Debbie Martyr's work in Sumatra, where she and her colleagues try to save habitat in general and the Sumatran tiger in particular. They also look for further evidence of the elusive orang-pendek. Membership begins at $22 and includes the magazine ***Flora & Fauna***. The group also publishes a leading journal of conservation, ***Oryx***.

Contact:
Flora and Fauna International - USA
Presidio Building 38, Suite 116
San Francisco, CA 94129-0156
Telephone: (800) 221-9524
Website: http://www.fauna-flora.org

One other American environmental group bears mentioning. Of the many organizations affiliated with the major U.S. political parties, Republicans for Environmental Protection (REP) is unique. REP tries to promote the conservation heritage of Theodore Roosevelt in a party often seen as overemphasizing economic growth at the expense of the environment. Visit at www.repamerica.org and decide for yourself.

NOTE ON TAXONOMY

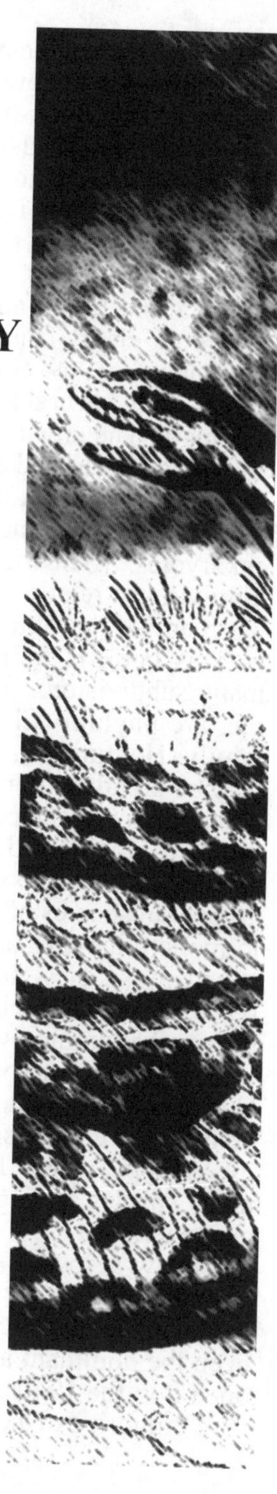

The frequent appearance of new species in scientific literature is sometimes due to new discoveries and sometimes to revised classifications. This book tries to avoid getting bogged down in technical details, but a basic understanding of taxonomy is important. Therefore, some readers may want a brief review.

The world of living organisms is divided into kingdoms. Originally, there were only two: plants and animals. The current consensus holds that there are five, with separate kingdoms for the bacteria, the fungi, and the protists (those microbes that don't fit anywhere else). This arrangement was first proposed by Robert Whittaker in 1969. Some experts think the protists should be split, perhaps into several kingdoms. Others add a higher division, called a domain. The creatures without organized cell nuclei, the bacteria and the archaea, make up two of the four domains, with viruses in the third. The fourth domain, the eukaryotes, includes all creatures made up of one or more nucleated cells. All plants and animals are eukaryotes. (Scientists are still debating what to do with the mimivirus, a relatively huge and complex virus found in 1992 that has many of the functions of a bacteria.)

This book is limited to the animal kingdom, which includes many subdivisions. The major ones, using humanity as our example, are phylum (Chordata), class (Mammalia), order (Primates), family (Hominidae), genus (Homo) and species (sapiens). All these may be broadened or narrowed using the prefixes super- and sub- : creating, for instance, superspecies and subspecies. This system produces the familiar binomial name incorporating genus and species, as in Homo sapiens or Latimeria chalumnae. As we are the "type" subspecies of Homo sapiens, it is proper (if clumsy) to refer to modern humans as Homo sapiens sapiens. In cases where there are many species, additional terms may be used to group related types, such as "species complex" and "tribe." These groupings are especially common when dealing with insects.

The basic building block of classification is the species. When Carl Linnaeus created the modern system of classification in the eighteenth century, species were differentiated based on morphology, or appearance. This turned out to be unworkable because so many species are highly variable in appearance and other physical characteristics.

Today, a species is defined as a population of animals which normally interbreed with each other and not with any other population. This idea, called the "biological species concept" (BSC), has been the dominant approach to taxonomy since Ernst Mayr developed it in the 1940s.

A minority of taxonomists, who believe the BSC focuses too

much on reproductive isolation, favor a more complex definition incorporating several factors or a definition based on the difference between the DNA of related populations. Given the current interest in the use of DNA analysis, it's important to note there is currently no universally accepted definition of what degree of variance in DNA indicates a distinction between species. That proper degree may vary considerably between different groups of animals, as the rate and type change in DNA is subject to numerous variables and assumptions. For purposes of this book, the BSC is more than adequate, and there is no need to go into the fine points of this debate.

Whatever definition is used, there is no single agreed-upon list of species. To offer just two examples: One genus of North American salamanders, Ensatina, was originally classified as one species but has now been split into four, and, in the opinion of some herpetologists, should have as many as eleven. As an example of reclassification in the other direction, Goeldi's monkey (*Callimico goeldii*) was once thought to merit its own family, but is now placed in the family Callitrichidae (the tamarins and marmosets).

Differences within a species may be recognized by dividing it into subspecies, although the delineation of subspecies is an imperfect art at best. Two other divisions sometimes used are "variety" and "geographic race." Some writers use these labels interchangeably with "subspecies," while others give them their own definitions. This is confusing even to biologists, some of whom advocate giving up on formal names below the species level. This book deals mainly with species, although subspecies are sometimes mentioned.

A final consideration is that, until the early twentieth century, scientists tended to be "splitters," creating new species based on very minor differences in morphology. When this led, for example, to 86 species of brown bears, it became obvious that the system was flawed. Today's zoologists who use the BSC are mostly "lumpers" who have reduced that swarm of bears to a single species with just four subspecies. Not surprisingly, there are also experts who think lumpers have sometimes gone too far.

The point of this book is not to split taxonomic hairs but to illustrate the richness and diversity of the animal kingdom. That is a point everyone can agree on, and one which becomes more apparent—and more wondrous—with every passing year.

much as reproductive isolation, favor a more complex definition that combines several factors, or a definition based on the differences between the DNA of isolated populations. With the current interest in the use of DNA analysis, it is important to note there is currently no universally accepted definition of what degree of variance in DNA [illegible] a distinct [illegible] species. That percentage may vary considerably between different groups of animals, the analysis and interpretation of DNA is subject to numerous variables and interpretation. For purposes of this book the BSC is more than adequate and there is no need to go into the fine points of this debate.

Whatever definition is used, the [illegible] of a single named [illegible] species, to offer just two examples: The [illegible] of North American salamanders [illegible] was originally described as one species, it has now been split into four, [illegible] the [illegible] biologists should [illegible] [illegible] [illegible] a [illegible] in the [illegible] [illegible] [illegible] [illegible] and [illegible] was once thought to be in its own family, but is now placed in the [illegible] [illegible] the [illegible] of [illegible].

[illegible] species [illegible] may be recognized [illegible] subspecies, although [illegible] [illegible] [illegible] [illegible] [illegible] [illegible] [illegible] [illegible] [illegible] variety and [illegible] have [illegible] [illegible] these terms interchangeably with "subspecies," while others give them their own [illegible]. There is [illegible] to botanists, some of whom [illegible] [illegible] [illegible] [illegible] [illegible] below the species level. This book deals mainly with species, although subspecies are sometimes mentioned.

[illegible] [illegible] [illegible] in the [illegible] [illegible] scientists [illegible] to be "splitters" [illegible] new species based on very minor differences [illegible] [illegible]. When [illegible], for example, [illegible] [illegible] [illegible] [illegible] [illegible] [illegible] that the system was [illegible]. Today's [illegible] who [illegible] the [illegible] [illegible] [illegible] have [illegible] that [illegible] of [illegible], of a single species [illegible] [illegible] [illegible] [illegible] [illegible] [illegible] are also [illegible] [illegible] [illegible] have sometimes [illegible] [illegible].

[illegible] [illegible] [illegible] to [illegible] [illegible] [illegible] the [illegible] and diversity of the animal kingdom. That is a point every one can [illegible], and [illegible] [illegible] more species [illegible] [illegible] with every passing year.

REFERENCES

SECTION I

Introduction

Ananova, 2004. "New species of shark discovered in Germany," http://www.ananova.com/news/story/sm_1073840.html, August.

Amman, Karl. 2003. "The Bondo Mystery Apes," at www.karlamman.com.

Gee, Henry. 2004. "Flores, God, and Cryptozoology," *Nature*, Published online October 27.

Jaeger, Peter. 2001. "A new species of *Heteropoda* (Araneae, Sparassidae, *Heteropodinae*) from Laos, the largest huntsman spider?" *Zoosystema*, 23 (3), p.461.

Mieszkowski, Katherine. 2002. "Data-mining life on earth," *Salon.com*, http://www.salon.com/tech/feature/2002/10/28/tree_of_life/index.html, October 28.

National Oceanic and Atmospheric Administration. 2001. "NOAA Investigates Giant Deep-Sea 'Mystery Squid,'" *NOAA Online News*, http://www.noaanews.noaa.gov/stories/s845.htm.

Roach, John. 2003. "Elusive African Apes: Giant Chimps or New Species?" *National Geographic News*, April 14.

Young, Emma. 2004. "The beast with no name," *New Scientist*, October 9, p.33.

New Guinea's New Kangaroo

Champkin, Julian. 1994. "Out of the trees, the kangaroo family's new branch," *Daily Mail* (London), July 21, p.19.

Flannery, Tim. 2003. Interview on radio program, *Enough Rope* with Andrew Denton, transcript online at http://www.abc.net.au/enoughrope/stories/s902427.htm.

Flannery, Tim. 1998. *Throwim Way Leg*. New York: Atlantic Monthly Press.

Flannery, Tim. 1995. *Mammals of New Guinea*. New York: Cornell University Press.

Fishing in South America

Anonymous. 2003. "A Biodiversity 'Scavenger Hunt,'" *Newsday*, November 11.

Anonymous. 1998. "Discovering New Fish Species," *Environmental News Network*, September 29. Available at http//www.enn.com/enn-multimedia-archive/1998/09/092998/newf.asp.

Conservation International. 2003. "New Fish Species Discovered in Venezuela," CI news release, August 28.

Conservation International. 1996. *CI News From the Front*, 2(2), October-November.
Keen, Cathy. 2000. "Last Unidentified Sport Fish in North America Gets a Scientific Name," University of Florida news release, January 20.
Lundberg, John. 1997. Personal communication, February.
Yoon, Carol Kaesuk. 1997. "Amazon's Depths Yield Strange New World of Unknown Fish," *New York Times*, February 12.

RAP Adventures

Anonymous. "Animal Info—Brasilia Burrowing Mouse, *Juscelinomys candango*," http://www.animalinfo.org/species/rodent/jusccand.htm.
Bowen, Lisa (Conservation International). 2000. Personal communications, February 28, March 2.
Conservation International (Various authors). News releases and papers, 1996-1999. All are available at www.conservation.org.
Cannell, Michael. 1999. "New Species—Keep on Counting," *Science World*, February 8.
Emmons, Louise. 2000. Personal communication, February 28.
Harrison, David. 2000. "New Mammal Discovered in South America," *The Sunday Telegraph* (London), February 28.
Nugroho, Kellk, and Rian Suryalibrata. 2000. "New Species Found in Irian Jaya," Pacific Islands Development Program, http://pidp.ewwc.hawaii.edu/PiReport/2000/October/10-24-17.htm.

How Many Mammals?

Anonymous. 2003. "New mammal hat trick," *BBC Wildlife*, January.
Anonymous. 2000. "DNA Research Reveals New Whale Species," *National Geographic News, http://news.nationalgeographic.com.*
Anonymous. 2000. "News Bytes" (discovery of new bat), *ENN News*, April 27.
Anonymous. 1999. "Stripy Bunny," *New Scientist*, August 21.
Anonymous. 1998. "Genetic Study Reveals Sumatran Tigers are a Distinct Species From All Mainland Tigers," *Science Daily*, June 10, http://www.sciencedaily.com.
Anonymous. 1998. "New Australian Marsupial Discovery," *The Australian*, May 22.
Dokuchaev, Nikolay. 1997. "A New Species of Shrew (Soricidae, *Insectivora*) From Alaska," *Journal of Mammology*, 78(3), p.811.
Eldridge, M. D. B., *et al.* 2001. "Taxonomy of rock-wallabies, *Petrogale* (Marsupalia: Macropodidae). III. Molecular data confirms the species status of the purple-necked rock-wallaby (*Petrogale purpureicollis* Le Souef)," *Australian Journal of Zoology*, 49, p.323.
Fair, James. 2003. "Borneo yields unlikely secret," *BBC Wildlife*, January.

Gonzales, Pedro, and Robert Kennedy. 1996. "A New Species of *Crateromys* (Rodentia: Muridae) From Panay, Philippines," *Journal of Mammology*, 77(1), p.25.
Jackson, Catriona. 2000. "Mystery Mammals Found - ANU Scientist," Canberra *Times*, September 5.
Morrell, Virginia, 1996. "New Mammals Discovered by Biology's New Explorers," *Science*, September 13, p.1491.
Roca, Alfred, *et al.* "Genetic Evidence for Two Species of Elephant in Africa," *Science*, August 24, p.1473.
Simpson, George. 1984. "Mammals and Cryptozoology," *Proceedings of the American Philosophical Society*, 128, p.1.
Tangley, Laura. 1997. "New Mammals in Town," *U.S. News and World Report*, June 9, p.59.
U.N. Wire, 2001. "New Species Discovered In China," February 6.
Williams, Robyn. 1999. Radio show transcript of interview with Dr. Colin Groves, November 12, http:www.abc.net/au/rn/science/ss/stories/s82699.htm.

The Still-Puzzling Coelacanth

Anonymous. 2000. "'Living Fossils' Discovered Off South Africa Coast," Reuters, December 1.
K. Hissmann, *et al.* 2000. "Conservation: Biogeography of the Indonesian coelacanths," *Nature,* Volume 403, Number 6765, p.38.
Anonymous. 1999. "A Second Living-Fossil Species?" *Science News*, April 24, p.267.
Anonymous. 1953. "Capture 'Living Fossil' Fish," *Science News Letter*, January 17.
Brown, George. 1995. Personal communication, April 10.
Brown, Malcolm. 1998. "Second Home of Fish From Dinosaur Age is Found," *New York Times*, September 24.
Courtney-Latimer, Marjorie. 1998. "The One That Got Away," *George*, December, p.72.
Erdmann, Mark. 2000. "New Home for 'Old Fourlegs,'" *California Wild*, Spring, p.8.
Holder, Mark, *et al.* 1999. "Two Living Species of Coelacanth?" *Proceedings of the National Academy of Sciences*, 96:22 (October), p.12616.
Erdmann, Mark, *et al.* 1998. "Indonesian 'king of the sea' discovered," *Nature*, 24 September, p.335.
Forey, Peter. 1998. "A home from home for coelacanths," *Nature*, 24 September, p.319.
Fowler, Stacey, and Shaher Misif. 1999. "The Living Fossil," *Environmental News Network*, http://www.enn.com/enn-features-archive/1999/12/121099/fossil_7046.asp.

Fricke, Hans, and Raphael Plante. 2001. "Silver coelacanths from Spain are not proof of a pre-scientific discovery," *Environmental Biology of Fishes*, 61 (4), p.463.
Greenwell, J. Richard. 1994. "Prehistoric fishing," *BBC Wildlife,* March.
McCabe, Heather, *et al.* 2000. "Tangled tale of a lost, stolen and disputed coelacanth," *Nature*, July 13, p.114.
Pouyaud, *et al.* 1999. "A new species of coelacanth," http://www.elsevier.fr/html/news/cras3mars99/pouyaud.html.
Shuker, Karl P. N. 1995. Personal communication, April 30.
Shuker, Karl P. N. 1998. Personal communication, October 22.
Thomson, Keith S. 1991. *Living Fossil: The Story of the Coelacanth*. New York: W.W. Norton.
Venter, Peter, *et al.* 2001. "Discovery of a viable population of Coelacanths (*Latimeria Chalumnae Smith*, 1939) at Sodwana Bay, South Africa," *Science in Africa*, Issue #2, November 22, http://www.scienceinafrica.co/za/coelnew.htm.
Weinberg, Samantha. 2000. *A Fish Caught in Time.* New York: HarperCollins.

The Mammals of Vu Quang

Anonymous. 2000. "Science Declares Rare Species a Bum Steer," *The Age* (Melbourne, Australia), December 18.
Adler, Hans George, 1995. "Antelope expose," *BBC Wildlife*, January, p.10.
Amato, George, *et. al.*, 1999. "A new species of muntjac, *Muntiacus putaoensis* (Artiodactyla: Cervidae) from northern Myanmar," *Animal Conservation*, 2, p.1.
Bauer, K. 1997. "Historic record and range extension for Giant Muntjac, *Muntiacus vuquangensis* (Cervidae)," *Mammalia*, 61(2), p.265.
Evans, T. D., *et al.* 2000. "Field observations of larger mammals in Laos, 1994-1995," *Mammalia*, 64(1), p.55.
Giao, P. N., *et al.* 1998. "Description of *Muntiacus truongsonensis*," *Animal Conservation*, 1, p.61.
Groves, Colin, and George Schaller. 2000. "The Phylogeny and Biogeography of the Newly Discovered Annamite Artiodactyls," in Vrba, Elisabeth S., and George B. Schaller (editors). 2000. *Antelopes, Deer, and Relatives: Fossil Record, Behavioral Ecology, Systematics, and Conservation*. New Haven, CT: Yale University Press.
Hassanin, *et al.* 2001. "Evidence from DNA that the mysterious 'linh duong' (*Pseudonovibos spiralis*) is not a new bovid," *Life Sciences* 324, p.71, available at http://www.elsevier.fr/html/news/VI-Hassanin.pdf.
Kuznetsov, G.V., *et. al.*, 2001. "The 'Linh Duong' *Pseudonovibos spiralis* (Mammalia, Artiodactyla) is a new buffalo," *Naturwissenschaften*, 88, p.123.

Line, Les. 1999. "Alan Rabinowitz; Indiana Jones Meets His Match in Burma Rabinowitz," *New York Times*, August 3.
Linden, Eugene. 1994. "Ancient Creatures in a Lost World," *TIME*, June 20, p.52.
Montgomery, Sy. 2002. *Search for the Golden Moon Bear*. New York: Simon and Schuster.
O'Rourke, Kate. 1999. "New Deer Species Found," *Discovery Online*, http://www.discovery.com/news/briefs/brief1.html.
Rabinowitz, Alan, 1997. "Lost World of the Annamites," *Natural History*, April 1997, p.14.
Rabinowitz, Alan. 1994. Personal communication, August 29.
Sautner, Stephen (WCS). 1999. Personal communication, July 2.
Shuker, Karl, 1999. Personal communication, July 19.
Shuker, Karl. 1995. "Vietnam—why scientists are stunned," *Wild About Animals*, March, p.32.
Thomas, Herbert, Arnoult Seveau, and Alexandre Hassanin. 2000. "The enigmatic new Indochinese bovid, *Pseudonovibos spiralis*: an extraordinary forgery," *Life Sciences* (Paris) 324, p.81.
Timm, Robert. 2000. Personal communication, December 18.
Timm, Robert, and John Brandt. 2000. "*Pseudonovibos spiralis* (Artiodactyla: Bovidae): new information on this enigmatic South-east Asian ox," *Journal of Zoology* (London), 253, p.157.
Van Dung, Vu, *et. al.*, 1993. "A New Species of Living Bovid from Vietnam," *Nature* 363, June 8, p.443.
Weiler, Hunter, *et. al.*, 1998. "The Distribution Of Tiger, Leopard, Elephant And Wild Cattle (Gaur, Banteng, Buffalo, Khting Vor and Kouprey) In Cambodia: Interim Report." Cambodian Wildlife Protection Office, July.
Wildlife Conservation Society, "Selected Papers from WCS Conservation Genetics Program," http://www.wcs.org.science.scienceresources/pubs.html.
World Conservation Monitoring Centre, "IUCN Red List Database," http://wcmc.org.uk.

Finding a Phylum

Angier, Natalie. 1995. "Flyspeck on Lobster Lips Turns Biology on its Ear," *New York Times*, December 14.
Anonymous. 1996. "Life on Lobster Lips," *Discover*, March.
Associated Press. 1995. "Tiny animal in a class by itself," Colorado Springs *Gazette-Telegraph*, December 17, p.A20.
Reuters. 2000. "Scientists Find Completely New Animal in Greenland," October 12.
Walker, Dave. 1996. "A Lobster's Microscopic Friend: *Symbion pandora* —a new life form and a new phylum," *Microscopy UK*,

http://www.microscopy-uk.org.uk/mag/articles/pandora.html.

Of Lemurs and Pottos

Anonymous. Golden-Brown Mouse Lemur (*Microcebus ravelobensis*), Primate Society of Great Britain, "Winter 1997 Meeting Report," Poster 5, December 3.

Anonymous. 1996. "Not a Potto," *Scientific American*, April.

Field Museum of Natural History. 2000. "Field Museum Announces New Discovery Made by Scientists Working in Madagascar," press release, November 13.

Friedman, Alan. 1997. "A Family Affair," *Pitt Magazine*, January.

Garbutt, Nick. 1999. *Mammals of Madagascar*. New Haven, CT: Yale University Press.

Handwerk, Brian. 2003. "Is This the Smallest Primate on Earth?" *National Geographic News*, http://news.nationalgeographic.com/news/2003/06/ 0626_030627_kingkong.html June 27.

Holden, Constance (editor). 1996. "Our New Relation," *Science*, March 1, p.1235.

Nelson, Bryn. 2004. "Madagascar to honor Stony Brook anthropologist," *Newsday*, June 4.

Rasoloarison, R.M., with Goodman and J.U. Ganzhorn. 2000. "Taxonomic revision of mouse lemurs (Microcebus) in the western portions of Madagascar," *International Journal of Primatology*. Vol. 21(6), p.963.

Wisconsin Regional Primate Research Center. 2000. "Golden-Brown Mouse Lemur (*Microcebus ravelobensis*)," http://www.primate.wisc.edu/pin/factsheets/microcebus_myoxinus.html, July 2, accessed September 8, 2000.

Zimmerman, E., *et al.* 1997. "A New Primate Species in Northwestern Madagascar: The

The Social Shrimp

Adler, Tina. 1996. "A shrimpy find: communal crustaceans," *Science News Online*, http://www.sciencenews.org, June 8.

Duffy, J. Emmett. 2003. "Underwater Urbanites," *Natural History*, December 2003-January 2004, p.40

Duffy, J. Emmett, *et al.* 1996-2000. Nine abstracts collected at *Abstracts of Recent Publications*, http://www.vims.edu/bio/mobee/abstract.htm #pub18.

MacDonald, Kirsti. 1999. *National Parks Journal* (Australia), October. Available at http://dazed.org/npa/npj/199910#shrimp.

Rosenberg, Yuval. 1999. "Predator Prawns," *Newsweek*, April 9.

South America's New Birds

Anonymous. 1999. "New antwren from the Western Amazonian lowlands," *World Birdwatch* 21(4), December.

Anonymous. 1999. "New species of brush-finch in Peru," *World Birdwatch* 21(2), June.

Anonymous. 1999. "Refuge Purchase for New Bird Species," *Birder's World*, February.

Astor, Michael, 1998. "New Bird Discovered in Brazil," Associated Press, April 24.

Astor, Michael. 1996. "New Bird Found, Nearly Extinct," Associated Press, November 16.

Brin, Dinah Weisenberg. 1998. "New Bird Discovered in Ecuador," Associated Press, June 10. de By, Rolf A. "Tapaculos," http://www.itc.nl/~deby/SM/NewTapaculos.htm.

Conservation International. 2003. "New Owl Species Discovered in Brazil," CI press release, June 12.

Graham, Rex. 1999. "In Search of Ecuador's Newest Bird," *Birder's World*, June, p.44.

Krabbe, Niels, *et al.* 1999. "A New Species of Antpitta (Formicariidae: *Grallaria*) From the Southern Ecuadorian Andes," *The Auk* 116(4), p.882.

O'Neill, John, *et al.* 2000. "A Striking New Species of Barbet (Capitoninae: *Captio*) From the Eastern Andes of Peru," *The Auk* 117(3), p.569.

Robbins, Mark, and F. Gary Stiles. 1999. "A New Species of Pygmy-Owl (Strigidae: *Glaucidium*) From the Pacific Slope of the Northern Andes," *The Auk* 116(2), p.305.

Whitney, Bret, and Jose Alvarez Alonso. 1998. "A New *Herpsilochmus* Antwren (Aves: *Thamnophilidae*) from Northern Amazonian Peru and Adjacent Ecuador," *The Auk*, 115(3), p.559.

World Wide Fund for Nature, *WWF News*, Summer 1996.

University of Kansas, 1999. "KU Museum Researcher Discovered New Pygmy-Owl Species," Press release, May 15.

The Smallest Frog

Bauer, Aaron. 1998. "Twentieth Century Amphibian and Reptile Discoveries," *Cryptozoology* (13), p.1.

Campbell, Jonathan. 2000. "A new species of venomous coral snake (Serpentes: *Elapidae*) from high desert in Puebla, Mexico," *Proceedings of the Biological Society of Washington*, 113(1), p.291.

Fountain, Henry, 1996. "Under Cuban Ferns, a Very Small Frog," *New York Times*, December 3.

Hanken, James. 1999. "4,780 and Counting," *Natural History*, July-August, p.82.

Holmes, Hannah. 1997. "The Lizard Wizard," *Wildlife Conservation*,

March/April, p.22.
Hoser, Raymond. 1998. "A New Snake From Queensland, Australia (Serpentes: *Elapidae*)," *Monitor,* 10(1), p.5.
Rocah, John. 2003. "Frog Discovery is 'Once in a Century,'" *National Geographic News,* http://news.nationalgeographic.com, October 15.
Slowinski, Joseph, and Wolfgang Wuster. 2000. "A New Cobra (Elapidae: *Naja*) From Myanmar (Burma)," *Herpetologica* 56(2), p.257.
Steitz, David. 2003. "NASA Helps Forecast Reptile Distributions in Madagascar," NASA Press Release 03-416, December 18.
Vergano, D., 1996. "Smallest frog leaps into the limelight," *Science News*, December 7, p.357.
Wilkinson, Paul. 1997. "Amateur Naturalist Discovers New Frog," *The Times* (London), June 5.
Wuster, Wulfgang. 2000. Personal communication, September 4.

The Carnivorous Sponge

Anonymous. 1995. "The Killer Sponge," *Discover*, July, p.22.
Kelly-Borges, Michelle. 1995. "Sponges out of their depth," *Nature*, January 26, p.284.
Stewart, Doug. "1999. "Sponges Get Respect," *International Wildlife*, July/August, p.26.
Vacelet, J., and N. Boury-Esnault. 1995. "Carnivorous sponges," *Nature*, January 26, p.333.
Vames, Steven. 1995. "Sponging Off Shrimp," *Scientific American*, May, p.18.

The Unknown Horses

Anonymous. 1992. "Wild Asian Horse to Return Home," *National Geographic*, March.
Anonymous. 1995. "Explorers Discover Previously Unknown Tibetan Pony," Reuters, November.
Day, David. 1990. *The Doomsday Book of Animals*. New York: Viking Press.
Groves, Colin P. 1974. *Horses, Asses, and Zebras in the Wild*. Hollywood, FL: Ralph Curtis Books.
Peck, Robert McCracken. 1999. "Home Again!" *International Wildlife*, September/October, p.36.
Peissel, Michael. 1997. *The Last Barbarians*. New York: Henry Holt.
Simons, Marlise. 1995. "A Stone-Age Horse Still Roams a Tibetan Plateau," *New York Times*, November 12.
Tsevegmid, D., and A. Dashdorj. 1974. "Wild Horses and other Endangered Wildlife in Mongolia," *Oryx*, February.

The New Galapagos

Miller, Mary. 1999. "Dispatches from Navassa" (series of articles),

http://www.discovery.com/exp/quest.dispatch.html.
Warrick, Joby. 1998. "Unsullied by Humans, U.S. Island is Biological Motherlode," *Washington Post*, August 17.

A Parade of Monkeys

Anonymous. 1998. "A New Brazilian Monkey," *Wildlife Conservation*, October, p.23.
Anonymous. 1997. "Scientist Finds New Monkey Species," Associated Press, August 18.
Anonymous. 1993. "Welcome to the Order," *Discover*, March.
Anonymous. 1991. "Stunning New Primate Species Found in Brazil," *National Geographic*, October.
Astor, Michael. 2000. "New Monkey Species Found in Brazil," Associated Press, April 22.
Burton, John A., and Bruce Pearson. 1987. *The Collins Guide to the Rare Mammals of the World*. Lexington, MA: The Stephen Greene Press.
Conservation International. 1996. "New Primate Species Discovered in Brazil," press release.
Ferrari, Stephen, and Helder Queiroz. 1994. "Two new Brazilian primates discovered, endangered," *Oryx*, January, p.31.
Goering, Laurie, 1999. "Amazon Primatologist Shakes Family Tree for New Monkeys," *Chicago Tribune*, July 11.
Huyghe, Patrick. 1993. "Remote Amazon gives old world new monkey," Associated Press, October 12.
Line, Les. 1996. "New Primate Species Discovered in Amazon Rainforest," *New York Times*, June 19.
Quammen, David. 2000. "The Rivers of Marmosets," *Whole Earth*, Fall, p.20.
Reilly, Patti. 1992. "500 Years Later, Scientists Continue to 'Discover' America: New Primate Species Found in Brazil," Conservation International, press release, October 13.
Rylands, Anthony, and Ernesto Luna (editors). 1996. "Discovery of a New Species of Marmoset in the Brazilian Amazon," *Neotropical Primates*, June.
Shuker, Karl. 2000. Personal communication, March 6.
Van Roosmalen, Marc. 2000. Personal communications, January 4 and March 2.

Van Roosmalen's Mammals

Anonymous. 1997. "Scientist Finds New Monkey Species," Associated Press, August 18.
Asato, Lani. 2000. "New Primates Discovered in Brazil's Forest," Conservation International press release, April 22.
Emmons, Louise. 1990. *Neotropical Rainforest Mammals*. Chicago:

University of Chicago Press.
Goering, Laurie. 1999. "Amazon Primatologist Shakes Family Tree for New Monkeys," *Chicago Tribune*, July 11.
McGirk, Tim. 2000. "A Rain-Forest Odyssey," *TIME*, February 28, p.75.
Shuker, Karl. 2003. *The Beasts That Hide From Man*. New York: Paraview.
Shuker, Karl. 2000. Personal communication, March 6.
Van Roosmalen, Marc. 2003. Descriptions on Web site, "New Species from Amazonia," http://www.amazonnewspecies.com/index.htm.
Van Roosmalen, Marc. 2000. Personal communications, January 4 and March 2.

Marine Life By the Numbers

Anonymous. 1997. "New Creatures Emerging From the Deep Sea," *CNN.com*, July 24.
Anonymous. 1996. "Count it Quick, Before It's Gone," *The Economist*, September 14, p.99.
Academy of Natural Sciences, 1999. "Ichthyologist Discovers a New Fish Species in the Waters of the South Pacific," March 30.
Arnaout, Rima. 2004. "International Mission to US Mid-Atlantic Finds New Species, Deep-Sea Mysteries," VOA news reported in Epoch Times, http://english.epochtimes.com/news/4-8-13/22850.html.
British Broadcasting Company, 2002. *The Blue Planet*, Programme 2, http://www.bbc.co.uk/nature/programmes/tv/blueplanet/programme2.shtml.
Broad, William. 1995. "The World's Deep, Cold Sea Floors Harbor a Riotous Diversity of Life," *New York Times*, October 17.
Dagit, Dominique Didier. 2003. Web page at http://clade.acnatsci.org/dagit/
De Forges, Bertrand Richer, *et al.* 2000. "Diversity and endemism of the benthic seamount fauna in the southwest Pacific," *Nature* (405) (June 22), p.944.
Ellis, Richard. 1996. *Deep Atlantic*. New York: Alfred A. Knopf.
Holden, Constance. 1997. "A New Denizen of the Deep?" *Science*, August 8, p.771.
Kettlewell, Julianna. 2003. "Ocean census discovers new fish," *BBC News Online*, October 23.
Lemonick, Michael. 1995. "The Last Frontier," *TIME*, August 14.
Macaulay, Craig. 2000. "Deep-Sea Scientists Find New Species on Sea Floor Near Australia," *National Geographic.com*, June 26.
Paxton, Charles. 1998. "A Cumulative Species Description Curve for Large Open Water Marine Animals," *Journal of the Marine Biological Association of the United Kingdom*, 78, p.1389.
Pilcher, Helen. 2004. "Four-armed jellyfish found," *BioEd Online*,

www.bioedonline.org, February 9.
Trivedi, Bijal. 2003. "New Jellyfish Species Found," *National Geographic News*, http://news.nationalgeographic.com, May 5.

The Amazing Octopus
Colin, Patrick. 1999. "Palau at Depth," *Ocean Realm*, Summer, p.77.
Norman, Mark. 2000. Personal communication, September 3.
Quammen, David. 2000. *The Boilerplate Rhino*. New York: Scribner.
Steene, Roger. 1998. *Coral Seas*. Buffalo, NY: Firefly Books.
Ross, John. 1999. "Masters of Mimicry," *Smithsonian*, March, p.112.
Turner, Pamela. 2003. "Uncommon Octopus," *Wildlife Conservation*, January, p.20.
Voight, Janet. 2000. Personal communication, August 17.

New Bird in New Zealand
BirdLife International, 2004. "Remarkable rail discovered 'just in time,'" www.birdlife.org.uk/news/news/2004/08/calayan_rail.html, August 17.
Dixon, Tina. 1997. "New Zealand Dog Team Helps Discover New Bird Species," *The Southland Times*, November 18.
Kennedy, Robert, *et al.* 1997. "New *Aethopyga* Sunbirds (Aves: Nectariniidae) From the Island of Mindanao, Philippines," *The Auk*, 114(1), p.1.
Muller, Karin. 2004. "New Zealand a Noah's Ark for Conserving Bizarre Birds," National Geographic News, September 21, http://news.nationalgeographic.com/news/2004/09/0921_040921_newzealand_birds.html.
Peterson, A. Townsend. 1998. "New Species and New Species Limits in Birds," *The Auk*, 115(3), p.555.
Rasmussen, Pamela. 1999. "A New Species of Hawk-Owl *Ninox* From North Sulawesi, Indonesia," *Wilson Bulletin*, 111, p.457. Cited in "Newly Described Species," http://www.dutchbirding.nl/pages/pagina_new.html. Other publications of new species by Rasmussen are cited in "Typical Owls—Recently described new species for the family," http://www.itc.nl/~deby/SM/NewTypicalOwls.htm.
Rasmussen, Pamela, *et al.* 2000. "A New Bush-Warbler (Sylviidae, *Bradypterus*) From Taiwan," *The Auk*, 117(2), p.279.
Taiwan Endemic Species Research Institute. No date. "Birds," http://www.tesri.gov.tw/content4/zool-e3.htm, accessed September 4, 2000.

Monitoring New Lizards
Anonymous. 1997. "At Home in the Rocks, a New Gecko Emerges," *National Geographic*, June.
Lemm, Jeff. 1998. "Year of the Monitor: A Look at Some Recently Discovered Varanids," *Reptiles*, September, p.70.
Reptile Exotics. No date. "Care Sheet—*Varanus melinus*,"

http://www.reptileexotics.com/melinus.html.
Sprackland, Robert. 1999. "A new species of monitor from Indonesia," *Reptile Hobbyist*, February.
Sweet, Samuel., and Eric Pianka. 2003. "The Lizard Kings," *Natural History*, November, p.40.

A New Turtle—Or Three

Anonymous. 2000. *The Vietnam Investment Review*, January 10.
Anonymous. 1998. "Giant Turtle Sightings Set Vietnam Capital Abuzz," *CNN.com*, April 13.
Anonymous. 1997. "Shell Game," *Discover*, January.
Duc, Ha Dinh. 2000. Email to Craig Heinselman, March 21.
Eliot, John. 1995. "New Tortoise Found After 30-Year Quest," *National Geographic*, August.
Heinselman, Craig. 2000. Personal communication, April 30.
McDonald, Mark. 1998. "Legendary Turtle Ready to Appear?" *The Record Online*, http://www.bergen.com/morenews/vietturtl199811197.htm.
Mason, Margie. 2003. "Group Moves to Save Endangered Turtle," Associated Press, November 3.
Von Radowitz, John. 2003. "Giant Vietnamese Turtle May Be Last of Breed," *The Scotsman*, October 12.

The Vu Quang Fish

Anonymous. 1996. Reuters dispatch, September 27.
Clover, Charles. 2000. "Valley Yields Creatures of a Lost World," London *Telegraph*, September 28.
Poston, Lee. 1996. "Vietnamese Scientist 'Reels In' a New Fish Species," WWF press release, September 27.
Royal Ontario Museum. 1999. "Dr. Bob discovers new species in Vietnam!" press release, July 8.
Royal Ontario Museum. 1995. "ROM Researchers Discover More New Species in Vietnam," press release, August.
Vietnam News Agency, 2004. "Over 130 new species of animal discovered in central national park," October 11.

Insects of the Past and Present

Anonymous. 2002. "Scientists discover 'giant bloated ants,'" *Ananova*. http://www.ananova.com, April 5.
Anonymous. 1998. "Berkeley Researchers Discover Tiny Moth," Environmental News Network, March 31. Available at http//www.enn.com/enn-news-archive/1998/03/033198/newmoth.asp.
Klass, Klaus, *et. al.*, 2002. "Mantophasmatodea: a new insect order with extant members in the Afrotropics," *Science*, May 24, 1456.

Mansfield, Duncan. 2000. "Moth Hunt Finds 700 Species," Associated Press, August 27.
Mulford, Kim. 2004. "C-P Reader Helps Rutgers Study by Catching Leeches," *Courier Post* (Camden, NJ), October 16.
Perlman, David. 1995. "Primitive Wasp Has Scientists Abuzz," *San Francisco Chronicle*, September 4, p.A15.

All the Species of the World

Anonymous. 2000. "Millennium Bug Found by Australian Entomologists," http:www.csiro.au, posted January 3.
Anonymous. 1997. "An Indecent Beetle," *Discover*, May, p.24.
Coniff, Richard. 1996. "What's in a Name? Sometimes More Than Meets the Eye," *Smithsonian*, December, p.66.
Howard, F.W. 1998. "How Many Insects Are There?" *Hort Digest* (1), November.
Isaak, Mark. 2000. "Curiosities of Biological Nomenclature," http://www.best.com/~atta/taxonomy.html.
Langreth, Robert. 1994. "The World According to Dan Janzen," *Popular Science*, December, p.78.
McCosker, John E. 2000. "What's in an Animal's Name?" *Whole Earth*, Fall, p.42.
Peterson, A. Townsend. 1998. "New Species and New Species Limits in Birds," *The Auk* 115(3), July, p.555.
Tangley, Laura. 1998. "How Many Species Exist?" *National Wildlife,* December 1998/January 1999, p.32.
Uetz, Peter. 2000. "How Many Reptile Species?" *Herpetological Review* 31(1), p.13.
United Nations Environment Programme. 1995. "UNEP Releases First Global Biodiversity Assessment Report," UN Press Release HE/916, November 14.
Wilson, Craig. 1994. "Presiding over a shrine to seashells," *USA Today*, September 28, p.1D.

Something New Out of Africa

Astor, Michael. 1998. "New Robin Species Found in Africa," Associated Press, August 21.
de By, Rolf A. "Recently Described Bird Species," http://www.itc.nl/~deby/SM/NewSpecies.html.
Safford, R.J., *et al.* 1995. "A New Species of Nightjar from Ethiopia," *Ibis* 137, 3, p.301.

The Weirdest Worms

Anonymous. 2000. "Worms in Gulf of Mexico Live L ong," Associated

Press, February 4.
Anonymous. 1997. "Scientists Discover Methane Ice Worms on Gulf of Mexico Sea Floor," Pennsylvania State University press release , July 29.
Anonymous. 1997. "Worms in Ice," *Earth*, December, p.14.
Pain, Stephanie. "1998. "Extreme Worms," *New Scientist*, July 25.

The Newest Whales

Anonymous. 2003. "Whale species is new to science," BBC News World Edition, http://news.bbc.co.uk, November 19.
Anonymous. 1991. "Shy Leviathan," *Discover*, November.
Australian Museum. 2003. "Longman's Beaked Whale," Fact sheet.
Baker, Mary L. 1987. *Whales, Dolphins, and Porpoises of the World*. Garden City, New York: Doubleday and Co.
Carwardine, Mark. 1995. *Whales, Dolphins and Porpoises*. New York: Doring Kindersley.
Dalebout, Merel. 2003. Personal communication, October 28.
Dalebout, Merel, *et al.* 2003. "Appearance, distribution, and genetic distinctiveness of Longman's beaked whale, Indopacetus pacificus," *Marine Mammal Science* 19:3, p.421.
Dalebout, Merel, *et al.* 2002. "A new species of beaked whale *Mesoplodon perrini* sp. n. (Cetacea: Ziphiidae) discovered through phylogenetic analyses of mitochondrial DNA sequences," *Marine Mammal Science* 18:3, p.577.
Ellis, Richard. 2003. Personal communication, November 22. Also 2000, March 10.
Ellis, Richard. 1980. *The Book of Wh*ales. New York: Alfred A. Knopf.
Forney, Karen. 2003. Personal communication, October 22. Also communications from 2000 and 1997, numerous dates.
Holmes, Bob. 2003. "New Whale Species Found in Museum," *New Scientist* news service, http://www.newscientist.com/news/news.jsp?id=ns99994402 November 18.
Leatherwood, Stephen, and Randall R. Reeves. 1983. *The Sierra Club Handbook of Whales and Dolphins*. San Francisco: Sierra Club Books.
Mead, James G., and Roger S. Payne. 1975. "A Specimen of the Tasman Beaked Whale, *Tasmacetus shepherdi*, From Argentina," *Journal of Mammology*, February.
Naish, Darren. 1998. Personal communication, December 16.
Nowack, Ronald M. 1991. *Walker's Mammals of the World*. Baltimore: Johns Hopkins University Press.
Ralls, Katherine, and Robert L. Brownell, Jr. 1991. "A whale of a new species," *Nature*, April 18.
Reyes, Julio C., *et al.* 1995. "*Mesoplodon bahamondi* sp.n. (Cetacea,

Ziphiidae), a New Living Beaked Whale From the Juan Fernandez Archipelago, Chile," *Bol. Museo Naional de Historia Natural* (45), Santiago, Chile, p.31.
Roach, James. 2003. "New Whale Species Announced by Japanese Scientists," *National Geographic News*. http://news.nationalgeographic.com/news/2003/11/1119_031119_rorqualwhale.html, November 19.
van Helden, A.L., *et al.* 2002. "Resurrection of *Mesoplodon traversii* (Grey, 1874), senior synonym of *M. bahamondi* Reyes, Van Waerebeek, Cárdenas and Yañez, 1995 (Cetacea: Ziphiidae). *Marine Mammal Science* (18), p.609.
Watkins, W.A. 1976. "A Probable Sighting of a live *Tasmacetus shepherdi* in New Zealand Waters," *Journal of Mammology*, May.
Urban-Ramirez, Jose. 1992. "First Record of the Pygmy Beaked Whale *Mesoplodon Peruvianus* in the North Pacific," *Marine Mammal Science,* October, p.420.
Wada, Shiro, *et al.* 2003. "A newly discovered species of living baleen whale," *Nature* 426, November 20, p.278.
Watson, Lyall. 1981. *Sea Guide to Whales of the World*. New York: E.P. Dutton.
Yamada, Tadasu. 2002. "On an unidentified beaked whale found stranded in Kagoshima," paper from the National Science Museum, Tokyo, December 25.

The Animals From Hell

Anonymous. 1999. "World's Hottest Worm," *Ocean Realm*, Summer, p15.
Childress, James, *et. al.*, "Symbiosis in the Deep Sea," in Gould, James L., and Carol Grant Gould, 1989. *Life at the Edge*. New York: W. H. Freemen and Company.
Gorman, Jessica. 2000. "Bubbling Under," *Discover*, March, p.12.
Kashefi, Kazem, and Derek Lovley. 2003. "Extending the Upper Temperature Limit for Life," *Science* (301), August 15, p.934.
Krajick, Kevin. 1999. "To Hell and Back," *Discover*, July, p.76.
Lemonick, Michael. 1995. "The Last Frontier," *TIME*, August 14.
Lutz, Richard. 2000. "Deep Sea Vents," *National Geographic*, October, p.116.
Marchant, Joanna. 2000. "Living it Up on Cloud Nine," *Space Daily*, August 28.
Pain, Stephanie. "1998. "Extreme Worms," *New Scientist*, July 25.
Recer, Paul. 2003. "Researchers find new world champion heat-tolerant microbe," Associated Press, August 15.
Tyson, Peter. 1999. "Neptune's Furnace." *Natural History*, June, p.42.

SECTION II

Introduction

Anonymous. 2000. "Recovery plan of the Hierro Giant Lizard," http://www.quercus.es/english/04PeptilesAnfibios/Gallotia.htm.

Associated Press. 2000. "African Monkey Pronounced Extinct," September 11.

IUCN. 2003 and 2004. "Red List," http://www.redlist.org.

IUCN. 2003. "Release of the 2003 IUCN Red List of Threatened Species," press release, November 18.

Kohn, David. 2004. "Here again, gone tomorrow?" March 8.

Louisiana Ornithological Society. 2000. "Ivory-Billed Woodpecker Sightings at Pearl River WMA?" *LOS News*, February.

Martel, Brent. 2000. "Birder Says He Saw Rare Woodpecker," Associated Press, November 4.

Quammen, David. 1998. "The weeds shall inherit the Earth," *The Independent* (London), November 22, p.30.

Mayell, Jillary. 2002. "'Extinct Woodpecker Still Elusive," *Naional Geographic News*, http://news.nationalgeographic.com, February 20.

Reid, W. R. 1992. "How many species will there be?" Chapter 3 in Whitmore, T.C., and J. A. Sayer (Editors), *Tropical Deforestation and Species Extinction*. New York: Chapman and Hall.

Shuker, Karl. 2003. "Ivory Bill Update," *Fortean Times*, August.

Reuters. 2000. "Spanish Scientists Discover 'Extinct' Lizards," March 18.

Search for the Woolly Flying Squirrel

Burton, John, and Bruce Pearson. 1987. *The Collins Guide to the Rare Mammals of the World.* Lexington, MA: Stephen Greene Press.

Yoon, Carol Kaesuk. 1995. "Woolly Flying Squirrel, Long Thought Extinct, Shows Up in Pakistan," *New York Times*, March 14.

Zahler, Peter. 1996. Personal communication, May 18.

Zahler, Peter. 1996. "Rediscovery of the Woolly Flying Squirrel (*Eupetaurus cinereus*)," *Journal of Mammology*, 77(1), p.54.

Who's That Owl?

Anonymous. 1996. *Conservation International News From the Front*, 2(4).

Associated Press. 1997. "Bird watchers spy rare owl for first time in 113 years," Colorado Springs *Gazette*, December 31.

Gallagher, Tim. 1998. "Lost and Found," *Living Bird*, Spring, p.24.

Hart, John. 1996. "Congo Bay Owl Rediscovered," *Wildlife Conservation*, October, p.10.

Kirby, Alex. 2003. "Fiji's 'extinct' bird flies anew," *BBC News Online*, November 28.

Kirby, Alex. 2002. "Parrots return after nine decades," *BBC News Online*, August 20.
Mountfort, Guy. 1988. *Rare Birds of the World*. London: Collins.

More Treasures From Southeast Asia

Amato, George, *et. al.*, 1999. "Rediscovery of Roosevelt's Barking Deer (*Muntiacus rooseveltorum*)," *Journal of Mammology*, 80, p.639.
Evans, T.D., *et al.* 2000. "Field observations of larger mammals in Laos, 1994-1995," *Mammalia*, 64(1), p.55.
Fisher, *et. al.*, IUCN, 1969. *Wildlife in Danger*. New York: Viking Press.
Groves, Colin P., et al. 1997. "Rediscovery of the wild pig *Sus bucculentis*," *Nature* (386), p.335.
Hebert, H. Josef, 1999. "Endangered Rhino Somehow Surviving," Associated Press, July 17.
Linden, Eugene. 1994. "Ancient Creatures in a Lost World," *TIME*, June 20, p.52.
Rabinowitz, Alan, 1997. "Lost World of the Annamites," *Natural History*, April 1997, p.14.
Shuker, Karl. 1995. "Vietnam—why scientists are stunned," *Wild About Animals*, March, p.32.

Gilbert's Potoroo

Anonymous. 1994. "Marsupial Thought Extinct is Found," *Dayton* (Ohio) *Dispatch*, December 8.
Day, David. 1989. *Vanished Species*. New York: Gallery Books.
Nowack, Rowland M. 1991. *Walker's Mammals of the World*. Baltimore: Johns Hopkins University Press.

Monk Seal Survival?

Adam, Peter, and Gabriela Garcia. 2003. "New information on the natural history, distribution, and skull size of the extinct (?) West Indian Monk Seal, *Monachus tropicalis*," *Marine Mammal Science*, 19:2, p.297.
Boyd, I.L., and M.P. Stanfield. 1998. "Circumstantial evidence for the presence of monk seals in the West Indies," *Oryx*, 32, p.310.
The Monachus Guardian (on-line journal) (2), http://www.monachus.org/mguard02/02mguard.htm.
Rice, Dale. 1998. *Marine Mammals of the World*. Lawrence, KS: The Society for Marine Mammology.
Swanson, Gail. 2000. "Final Millennium for the Caribbean Monk Seal," *The Monachus Guardian* 3(1), http://www.monachus.org/mguard05/ 05infocu.htm.
Walters, Mark. 1997. "Ghost of a Monk Seal," *Animals*, November/December, p.23.

Birding in Indonesia

Anonymous. 2000. "'Extinct'" Birds Found on Islet," Agence France-Presse report, July, posted at http://www.hkbws.org.hk/tempweb.html.

Anonymous. 2000. "Chinese seabird 'returns from extinction,'" British Broadcasting Corporation report, July 26.

Anonymous. 1996. "Rare bird species winging way back," *Colorado Springs Gazette-Telegraph,* February 23, p.E1.

Irwin, Aisling. 1999. "Britons find 'extinct' blue flycatcher," *Daily Telegraph* (England), August 30.

Mountfort, Guy. 1988. *Rare Birds of the World.* London: William Collins and Sons.

World Nature Network. No date. "Chinese Crested Tern," http://www.wnn.or.jp/wnn-asia/a_bird_e/English/Bird_to_Watch/3247_tern.html.

World Nature Network. No date. "Lompobattang Flycatcher," http://www.wnn.or.jp/wnn-asia/a_bird_e/English/Bird_to_Watch/6497_flycatcher.html.

Lemur Resurrection

Anonymous. 1995. "Return of the little lemur," *New Scientist*, 15 July, p.11.

Wisconsin Regional Primate Research Center. 2000. "Pygmy Mouse Lemur (*Microcebus myoxinus*)," http://www.primate.wisc.edu/pin/factsheets/microcebus_ravelobensis.html, July 2.

Is the Deer Still There?

Byun, S.C., *et. al.*, 2002. "Evolution of the Dawson caribou (*Rangifer tarandus dawsoni*)," *Canadian Journal of Zoology* Vol. 80, p.956.

Ceska, Adolf, 1995. Posting on Biodiversity listserve concerning the type specimen of Dawson's caribou, December 23, http://biodiversity.uno.edu /~gophtax/_gophtax.95/0774.html.

Day, David. 1989. *Vanished Species*. New York: Gallery Books.

Environment Canada, *Species at Risk*, "Caribou dawsoni subspecies," http://www.speciesatrisk.gc.ca/search/speciesDetails_e.cfm?SpeciesID=7

IUCN. 2004. Red List of Threatened Specie, "Cervus schomburgki," available at http://www.redlist.org/search/details.php?species=4288.

Kingsbury, Bruce, n.d. "On the Mechanism of Genetic Drift—understanding (1-1/2Ne), http://users.ipfw.edu/kingsbury/COURSES/Conservation/RIFT.htm.

MacPhee, R.D.E. and Flemming, C. 1999. "Requiem Æternam, The last five hundred years of mammalian species extinctions." In: R.D.E. MacPhee (ed.), *Extinctions in Near Time*, pp.333-371. Kluwer Academic/Plenum Publishers, New York.

Schroering, George. 1991. "Swamp Deer Resurfaces," *Wildlife Conservation*, December, p.22.

Bringing Back the Dead

Anonymous. 2000. "Extinct' Lions Surface in Siberia," BBC, November 5.
Anonymous. 2000. Associated Press, January 10.
Anonymous. 1999. "Ancient Urchins Reborn," *Ocean Realm*, Summer, p.15.
Anonymous. 1999. "Tasmanian Tiger Clone 'Impossible,'" *Daily Telegraph* (Australia), September 28.
Anonymous. 1999. "Cloning of extinct Huia bird approved," Environmental News Network, http://www.enn.com, July 20.
Associated Press. 1997. "Mammoth Sperm Search Fails," September 19.
Boroughs, Don. 1999. "New Life for a Vanished Zebra?" *International Wildlife*, March/April, p.46.
Burton, John, and Bruce Pearson. 1988. *The Collins Guide to the Rare Mammals of the World*. Lexington, MA: Stephen Greene Press.
Flaccus, Gillian. 2002. "Couple tries to re-create extinct horse breed," Associated Press, June 27.
Greenway, James C., Jr. 1967. *Extinct and Vanishing Birds of the World*. New York: Dover Publications.
Harbury, Martin. 1984. *The Last of the Wild Horses*. Garden City, NY: Doubleday.
Harris, Paul. 1998. "Extinct Zebras May be Back," Associated Press, August 9.
Hopkins, Andrea. 2000. "Clone That Tiger? It Could Happen," Reuters, August 21.
McNeil, Donald. 1999. "Barbary lion found in abandoned traveling circus proves rare," Denver *Rocky Mountain News*, June 28, p.2A.
McNeil, Donald. 1997. "Brave Quest of Africa Hunt: Bringing Back Extinct Quagga," *New York Times*, September 16.
Perlman, Heidi. 2000. "Scientists Close to Cloning Endangered Animals," Associated Press, October 8.
Shuker, Karl. 1997. *From Flying Toads to Snakes With Wings*. St. Paul, MN: Llewellyn Publications.
Stone, Richard. 1999. "Cloning the Woolly Mammoth," *Discover*, April, p.56.
Weiss, Rick. 1998. "New Zealand Scientists Clone the Last Cow of a Rare Breed," *Washington Post*, August 20.

The Thylacine—Dead or Alive?

Anonymous. 1999. "Tasmanian Tiger Could be Brought Back From the Dead," *The Advertiser* (Adelaide), May 13.
Anonymous. 1997. *Daily Telegraph* (Australia). April 4, p.17.
Ashley-Griffiths, Katy. 1997. "A New Hunt for "Extinct' Tiger," *Herald and Weekly Times*, October 26.
Drollette, Daniel. 1996. "On the Trail of the Tiger," *Scientific American*, October, p.32.

Guiler, Eric, and Phillipe Goddard. 1998. *Tasmanian Tiger: A Lesson to be Learnt*. Perth, Australia: Abrolos Publishing.
Healy, Tony, and Paul Cropper. 1994. *Out of the Shadows*. Chippendale, Australia: Ironbark.
Reuters. 1997. Wire service dispatch, March 24.
Sutherland, Struan. 1995. "Sighting put bite in tiger tale," *Australian Doctor*, March 31.
Tasmania Parks and Wildlife Service. "Thylacine, or Tasmanian Tiger, *Thylacinus cynocephalus*," fact sheet available at http://www.parks.tas.gov.au/wildlife/mammals/thylacin.html.
Williams, Vanessa. 1999. "Rogue 'puma' returns," *Herald Sun*, November 26.

The Cat Came Back

Anonymous. 2000. "Police 'big cat' warning," British Broadcasting Co. report, August 25.
Armstrong, Adam. 2000. "Beware of the big cats," *The Times* (London), September 5.
Bolgiano, Chris. 1995. *Mountain Lion*. Mechanicsburg, PA: Stackpole Books.
Bolgiano, Chris. 2000. Personal communication (via Loren Coleman). August.
Coleman, Loren. 2000. Personal communications, August 14, August 26.
Downing, Robert L. 1982. *Eastern Cougar Recovery Plan*. Atlanta, GA: U.S. Fish and Wildlife Service.
Ewing, Susan, and Elizabeth Grossman (editors). 1999. *ShadowCat*. Seattle: Sasquatch Books.
Grahnke, Lon. 2000. "First State Mountain Lion Since 1862 Found Dead," Chicago *Sun Times*, July 19.
Lefkowitz, Melanie. 1998. "Mountain Lion Observed Again in Rhode Island," Providence *Journal*, December 29.
Lowe, David, et al. 1990. *The Official World Wildlife Fund Guide to Endangered Species of North America*. Washington, D.C.: Beacham.
McNamee, Thomas. 1980. "Chasing a Ghost," *Audubon*, March.
Sharp, Eric. 2001. "Cougars Still Prowl in Michigan," *Free Press*, November 1.
Shuker, Karl P.N. 1989. *Mystery Cats of the World*. London: Robert Hale.
Tinsley, Jim Bob. 1987. *The Puma*. El Paso: Texas Western Press.
U.S. Fish and Wildlife Service. 1992 (?) "Eastern Cougar," (fact sheet), http://endangered.fws.gov/i/a/saa48.html.
Wright, Bruce S. 1961. "The Latest Specimen of the Eastern Puma," *Journal of Mammology*, May.

The Howling God

Anonymous. 1994. "Search (for?) the Japanese Wolf," *The Nihon Keizai*

Shimbun, February 15.
Anonymous. 1994. "The Japanese Wolf Surely Survives," *The Nihon Keizai Shimbun,* March 20.
Anonymous. 2000. "Canid Resembling Extinct Japanese Wolf Sighted," *Yomiuri Shimbun*, November 20.
Day, David. 1989. *Vanished Species*. New York: Gallery Books.
Knight, John. 1997. "On the Extinction of the Japanese Wolf," *Asian Folklore Studies* (56), p.129.
Shuker, Karl. 1991. *Extraordinary Animals Worldwide*. London: Robert Hale.
Takakura, Tomoaki. 2000. "'Extinct' wolf may still be on the prowl," *Mainichi Shimbun*, November 22.
United States Fish and Wildlife Service. No date. "Red Wolf (*Canis rufus*)," fact sheet, http://species.fws.gov/bio_rwol.html.
Wayne, Bob. 1995. "Red Wolves: to Conserve or not to Conserve," CANID NEWS, IUCN Canid Specialist Group, Vol. 3, 1995
Yanai Kenji, 1993. "Visionary Japanese Wolves," personal account in unknown publication, March.

NOTE: Some of the Japanese publications cited here were translated into English by Ishizawa Naoya and provided to me by Angel Morant Fores.

SECTION III

Introduction

Carroll, Todd. 1998. *The Skeptic's Dictionary*, http://skeptic.com/ crypto.html.
Coleman, Loren, and Jerome Clark. 1999. *Cryptozoology A to Z*. New York: Fireside.
Heuvelmans, Bernard. 1982. "What is Cryptozoology?" *Cryptozoology* (1), p.1.

The British Naturalists' Sea Monster

Ellis, Richard. 2003. *Sea Dragons*. Lawrence, Kansas: University Press of Kansas.
Ellis, Richard. 1998. *The Search for the Giant Squid*. New York: Lyons Press.
Ellis, Richard. 1994. *Monsters of the Sea*. New York: Knopf.
Gould, Rupert T. 1930. *The Case for the Sea Serpent*. London: Philip Allan.
Harrison, Paul. 2001. *Sea Serpents and Lake Monsters of the British Isles*. London: Robert Hale.
Heuvelmans, Bernard. 1968. *In the Wake of the Sea Serpents*. New York: Hill and Wang.
Meade-Waldo, E.G.B., and Nicoll, Michael J., 1906. "Description of an Unknown Animal Seen at Sea off the Coast of Brazil," *Proceedings of*

the Zoological Society of London, p.719.
Nicoll, Michael J. 1908. *Three Voyages of a Naturalist*. London: Witherby and Co.
Molloy, R. 1915. "A Queer Tale of Flanagan and the Eel off Dalkey Sound," publication title unknown, August 28. Available at http://www.clubi.ie/dalkeyhomepage/ee.html.
Taylor, L.R., Compagno, L.J.V., and Struhsaker, P.J. (1983). "Megamouth—a new species, genus, and family of lamnoid shark (*Megachasma pelagios*, family Megachasmidae) from the Hawaiian Islands," *Proceedings of the California Academy of Sciences, vol.* 43, p.87.

The Primate Problem

Anonymous. 1982. "Big Foot Fraud," *OMNI*, September.
Coleman, Loren, and Patrick Huyghe. 1999. *The Field Guide to Bigfoot, Yeti, and Other Mystery Primates Worldwide*. New York: Avon.
Coleman, Loren. 1999. *Cryptozoology A to Z*. New York: Fireside.
Dalton, Rex. 2004. "Little lady of Flores forces rethink of human evolution," *Nature*, Published online October 27.
Davis, Anthony. 2003. "Searching for Sasquatch," *Texarkana* (TX) *Gazette*, October 20.
Gee, Henry. 2004. "Flores, God, and Cryptozoology," *Nature*, Published online October 27.
Hocking, Peter. 1992. "Large Peruvian Mammals Unknown to Zoology," *Cryptozoology* (11), p.38.
Napier, John. 1972. *Bigfoot*. New York: Berkeley.
Sanderson, Ivan. 1969. "The Missing Link?" *Argosy*, May, p.23.
Sanderson, Ivan. 1961. *Abominable Snowmen*. Philadelphia: Chilton.
Shackley, Myra. 1983. *Still Living?* New York: Thames and Hudson.
Wylie, Kevin. 1980. *Bigfoot*. New York: Viking.
Meryhew, Richard. 2000. "Old tractor up for sale, reluctantly, by owner," Minneapolis *Star Tribune*, September 23, p.A1. (Article on Frank Hansen)

Is the Yeti Still Out There?

Beeby, Adrian. 2000. "Stalking Yetis at the World's Edge," London *Evening Standard*, January 17.
Ciochon, Russell. 1996. Personal communication, August 17.
Ciochon, Russell, *et al.* 1990. *Other Origins*. New York: Bantam.
Coleman, Loren. 2003. Personal communication, January 13: also 2001, October 15.
Coleman, Loren. 1989. *Tom Slick and the Search for the Yeti*. Boston: Faber and Faber.
Crossland, David. 1998. "Mountaineer Muddies Myth of the 'Yeti,," Reuters, October.

Krantz, Grover. 1987. "A Reconstruction of the Skull of *Gigantopithecus blacki* and its comparison with a Living Form," *Cryptozoology* (6), p.24.
Lorenzi, Rosella. 2001. "Scientists Claim Yeti DNA Evidence," *Discovery News*, April 2.
Napier, John. 1972. *Bigfoot.* New York: Berkeley.
Nickell, Joe. 1995. *Entities*. Buffalo, NY: Prometheus Books.
Tschernezky, Wladimir. 1960. "A Reconstruction of the Foot of the 'Abominable Snowman," *Nature*, May 7, p.496.
Zahorka, Herwig. 1997. Letter, Jakarta *Post*, August 8.

Bye-Bye, Bigfoot?

Ashton, Linda. 1999. "Man Claims to be Bigfoot in Famous 1967 Sasquatch Film," *The Oregonian*, February 2.
Coleman, Loren. 2004. "Costume Drama," *Fortean Times*, August, p.58
Coleman, Loren. 2002. Personal communications, December 7, 13: Also 2000, October 15.
Green, John. 1998. Personal communication to Loren Coleman, August 31.
Krantz, Grover. 2000. Personal communication, February 1.
Krantz, Grover. 1999. *Bigfoot Sasquatch Evidence.* Blaine, WA: Hancock House.
Long, Greg, and Karl Korff. 2004. *The Making of Bigfoot*. New York: Prometheus. 475pp.
Loxton, Daniel. 2004. "Big Foot Con," *Skeptic*, Vol. 11, No. 1, p.82.
Meldrum, Jeff, and Richard Greenwell. 1998. "Bigfoot: take two." *BBC Wildlife*, September.
Napier, John. 1972. *Bigfoot.* New York: Berkley
Packham, Chris. 1998. "Bigfoot: Proof or Spoof?" *BBC Wildlife*, September.
Stein, Theo. 2003. "Bigfoot Believers," *Denver Post*, January 5, p.1A.

The Enduring "Sea Serpent"

Anonymous. 1930. No title (report of the HMS *Hilary* sea serpent), *Nature* (125), March 22, p.469.
Anonymous. 2003. "Reward out for 'Milford Monster' spotted by pub's lunch customers," *Western Mail* (Wales), March 13.
Bauchot, Roland (Editor). 1999. *Snakes: A Natural History*. New York: Sterling Publishing Co.
Bauer, Aaron, and Anthony Russell. 1996. "A Living Plesiosaur? A Critical Assessment of the Description of *Cadborosaurus willsi*," *Cryptozoology* (12), p.1.
Benchley, Peter. 2002. *Shark Trouble*. New York: Random House.
Bright, Michael. 1989. *There are Giants in the Sea.* London: Robson Books.

Carrington, Richard. 1957. "Sea Serpent—Riddle of the Deep," *Natural History*, April, p.183.

Champagne, Bruce. 2003. Personal communications, November and December (several dates).

Champagne, Bruce. 2002. "A Preliminary Evaluation of a Study of the Morphology, Behavior, Autoecology, and Habitat of Large, Unidentified Marine Animals," *Dracontology* (1), (specialty publication edited and printed by Craig Heinselman, Francestown, New Hampshire).

Corliss, William. 1984. "California 'Sea Serpent' Flap," *Science Frontiers*, May-June.

Drummond, Maldwin. 1995. Correspondence, February 28.

Ellis, Richard. 1994. *Monsters of the Sea*. New York: Knopf.

Frizzell, M.A. "'Chessie' (The Chesapeake Bay Phenomenon)," http://umbc7.umbc.edu/~frizzell/cryptozoo.html.

Gardener, Matt Hunt. 2003. "C.B. lobster fisherman follows 'sea monster,'" *Halifax Herald*, June 25.

Gilmore, C.W. 1938. "Fossil Snakes of North America," Geological Society of America, Special Papers (9).

Gordon, David. 1987. "What is That?" *Oceans*, August, p.44.

Gould, Rupert T. 1930. *The Case for the Sea Serpent*. London: Philip Allan.

Gustafson, Bob. 2003. "'Sea serpent' surprises Nova Scotia lobsterman," *Working Waterfront / Inter-Island News*, August, p.3.

Harrison, Paul. 2001. *Sea Serpents and Lake Monsters of the British Isles*. London: Robert Hale.

Heuvelmans, Bernard. 1968. *In the Wake of the Sea Serpents*. New York: Hill and Wang.

LeBlond, Paul H. 2001. "Earliest Cadborosaurus Sighting from 1791 Reported," *BCSCC Quarterly*, Spring, p.4.

LeBlond, Paul H., and Edward Bousfield. 1995. *Cadborosaurus: Survivor From the Deep*. Victoria, B.C.: Horsdal and Schubart.

LeBlond and Bousfield. 1995. "An Account of *Cadborosaurus willsi*, new genus, new species, a large aquatic reptile from the Pacific coast of North America," *Amphipacifica* (I), Supplement I, April 20.

Ley, Willy. 1959. *Exotic Zoology*. New York: Viking.

Ley, Willy. 1957. "The Great Sea-Serpent Mystery," *Science Digest*, February, p.6.

Nicoll, Michael J. 1908. *Three Voyages of a Naturalist*. London: Witherby and Co.

Rubenstein, Steve. 1983. "Sea Serpent Captured on Paper," *San Francisco Chronicle*, November 3.

Sundberg, Jan. 2003. "The Cornwall 'plesiosaur' is a bird," http://www.bahnhof.se/~wizard/cryptoworld/index219a.html.

Whiffen, Glen. 2000. "Sea monster sighted off Bonavista," *The Telegram* (St. John's, Newfoundland), April 6.
Wood, J.G. 1884. "The Trail of the Sea-Serpent," *Atlantic Monthly*, June, p.799.
"R.L.," 1906. "The Sea-Serpent," *Nature* (74), June 28, p.202.

Loch Ness: What's Going On?

Anonymous. 1998. "Cameraman fuels new 'Nessie' speculation," Reuters, March 2.
Bauer, Henry. 2002. "Common Knowledge about the Loch Ness Monster: Television, Videos, and Films," *Journal of Scientific Exploration* (16:3), p.455.
Bauer, Henry. 2002. "The Case for the Loch Ness 'Monster': The Scientific Evidence," *Journal of Scientific Exploration* (16:2), p.225.
Bauer, Henry. 1991. *The Enigma of Loch Ness.* Johnston and Bacon.
Coleman, Loren. 2000. Personal correspondence, March 5.
Coleman, Loren. 1998. Personal correspondence, February 15.
Campbell, Steuart. 1991. *The Loch Ness Monster: The Evidence.* Aberdeen University Press.
Greenwell, Richard. 1988. *The ISC Newsletter*, Spring.
Gould, Rupert. 1969 (reprint of 1934 original). *The Loch Ness Monster and Others.* Secaucus, NJ: Citadel Press.
Hughes, Lorna. 2000. "It Couldn't be Nessie…Could it?" *Sunday Mail* (Scotland), October 29.
Mackal, Roy. 1980. *The Monsters of Loch Ness*. Chicago: Swallow Press.
Owen, James. 2003. "Loch Ness Sea Monster Fossil a Hoax, Say Scientists," *National Geographic News*, http://news.nationalgeographic.com, July 29.
Raynal, Michel. 2003. Post to newsgroup cryptolist@yahoogroups.com, July 31.
Raynor, Dick. 2000. Personal communication, October 7.
Razdan, Rikki, and Alan Kielar. 1984. "Sonar and Photographic Searches for the Loch Ness Monster: A Reassessment," *Skeptical Inquirer* (9) (Winter), p.147.
Sheldon, R., and S. Kerr, 1972. "The Population Density of Monsters of Loch Ness," *Limnology and Oceanography*, (17), p.796.
Trottman, Andreas. 2001. "Andreas Trottman Reports from Loch Ness," *BCSCC Quarterly*, Spring, p.6.
Young, Noel. 1998. "U.S. Search for Nessie Finds Moby Dick," *Scotland on Sunday*, January 18.

The State of the Sasquatch

Anonymous. 1995. "New Light on Bigfoot—a Legendary Hoax?" *Daily*

Telegraph (London), April 24.
Aldridge, Dorothy. 1978. "Endangered Species Paper Plugs Bigfoot Hunters' Guns," *Colorado Springs Gazette-Telegraph*, January 25.
Begley, Sharon. 1987. "Tracking the Sasquatch," *Newsweek*, September 21, p.73.
Coleman, Loren. 2002. Personal communications, December 7, 13.
Cote, Suzanne. 2004. "Origins of the African hominoids: an assessment of the palaeobiogeographical evidence," *Comptes Rendus Palevol*, No. 3, p.321.
della Cava, Marco. 2002. "Bigfoot's indelible imprint," *USA Today*, October 31.
Fay, John (U.S. Fish and Wildlife Service). 1988. Personal communication, November 22.
Freeman, Mark. 2000. "The Man Who Saw Bigfoot," *Wildlife Journal*, Winter, p.68.
Gallagher, Dan. 1999. "Idaho Professor on the Trail of Bigfoot," *The Deseret News*, May 24.
Goodall, Jane. 2002. Interview with National Public Radio, September 27.
Goodavage, Maria. 1996. "Search for Sasquatch Becomes a Bear of a Task," Detroit *News*, May 28.
Greenwell, J. Richard, and James E. King. 1981. "Attitudes of Physical Anthropologists Toward Reports of Bigfoot and Nessie," *Current Anthropology,* February, p.79.
Kirk, John. 1999. "Patterson film survives assaults on its authenticity," *CryptoNews*, January, p.5.
Kleiner, Kurt. 2000. "Bigfoot's Buttocks," *New Scientist Weekly Newsletter*, 23/20 December, http://www.newscientist.com/news/news.jsp?id=ns227015.
Krantz, Grover. 2000. Personal communication, February 1.
Krantz, Grover. 1999. *Bigfoot Sasquatch Evidence*. Blaine, WA: Hancock House.
Locke, Michelle. 2000. "Mountain Village Has Big Plans for Bigfoot," Seattle *Post-Intelligencer*, July 17.
Lore, David. 1995. "DNA scientists at OSU on a quest for Bigfoot," *Columbus Dispatch*, November 3.
McCafferty, Keith. 2000. "Print Pro Says Bigfoot May Exist," *Field & Stream*, January, p.15.
Napier, John. 1972. *Bigfoot.* New York: Berkley.
Poirier, Frank (Ohio State University). 1998. Personal communication, January 11.
Pyle, Robert. 1995. *Where Bigfoot Walks*. Boston: Houghton Mifflin.
Starr, Douglas. 1982. "Big Foot Fraud," *OMNI*, September, p.100.
Stein, Theo. 2003. "Bigfoot Believers," Denver *Post*, January 5, p.1A.
Wylie, Kenneth. 1980. *Bigfoot*. New York: Viking Press.

Weintraub, Pamela. 1985. *OMNI's Catalog of the Bizarre.* Garden City, NY: Doubleday.

Creatures of the Lakes

Anonymous. 2000. "Reward Offer for Nessie's Cousin," *South China Morning Post*, May 3.
Cockerham, William. 1981. "Vermonters Defend Lake 'Champ,'" *Hartford Courier*, August 31.
Coleman, Loren. 2000-2003. Personal communications, numerous dates.
Deuel, Richard, and Dennis Hall. 1992. "Champ Quest at Lake Champlain, 1991-1992," *Cryptozoology* (11), p.102.
Gaal, Arlene. 2003. Personal communication, October 23. Also 2000, March 5.
Kirk, John. 1999. Personal communications, numerous dates.
Kirk, John. 1998. *In the Domain of the Lake Monsters*. Toronto: Key Porter Books.
Lehn, W.H. 1979. "Atmospheric Refraction and Lake Monsters," *Science*, July 13, p.183.
Mackal, Roy. 1980. *Searching for Hidden Animals*. New York: Doubleday.
Moon, Mary. 1978. "Ogopogo, Canada's lake monster: oft seen, never snared," *Smithsonian*, November, p.173.
Naish, Darren. 1997. "Ancient Whales, Sea Serpents and Nessies, Part 2: Theorising on Survival," *Animals & Men* (10), p.13.
Nickell, Joe. 2003. "Legend of the Lake Champlain Monster," *Skeptical Inquirer*, July.
Parsons, Serena. 2004. "Family videos what could be Ogopogo," *The Daily Courier* (Kelowna, B.C.), October 22.
Teresi, Dick. 1998. "Monster of the Tub," *Discover*, April, p.86.
Vachon, Brian. 1978. "Is There a Champlain Monster?" *Reader's Digest*, April, p.9.
Zarzynski, Joseph. 1993. "'Champ' A Zoological Jigsaw Puzzle," *Adirondack Bits 'n Pieces* (1:1).
Zarzynski, Joseph. 1986. *Monster Wrecks of Loch Ness and Lake Champlain*. Wilton, NY: M-Z Information.

Protecting Nessie

Anonymous. 1998. *Guardian* (Editorial), January 1.
Balakrishnan, Peter. 1998. "Monsters from the Hoary Past," *Business Standard*, January 1.
Harvey, Amanda. 2001. "All Nessie-sary Precautions," The *Scotsman*, June 9.

SECTION IV

Peruvian Mystery Cats

Emmons, Louise. 2000. Personal communication, February 28.

Hocking, Peter. 2000. Personal communication, September 12. Also 1999 and 1997, various dates.
Hocking. 1996. "Further Investigation Into Unknown Peruvian Mammals," *Cryptozoology* (12), p.50.
Naish, Darren. 2000. Personal communication, October 24.
Shuker, Karl. 2003. *The Beasts That Hide From Man*. New York: Paraview.

Quest for the Golden Bear

Bille, Matthew. 1995. *Rumors of Existence*. Blaine, WA: Hancock House.
Day, David. 1990. *The Doomsday Book of Animals*. New York: Viking Press.
Domico, Terry. 1988. *Bears of the World*. New York: Facts on File.
Goodwin, George. 1946. "Inopinatus the Unexpected," *Natural History*, November.
Halfpenny, James. 1996. "Tracking the Great Bear: Mystery Bears," *Bears*, Spring.
Montgomery, Sy. 2003. *Search for the Golden Moon Bear*. New York: Simon and Schuster.

Requiem for the Giant Octopus?

Anonymous. "AO-34 USS Chicopee," http://metalab.unc.edu/hyper-war/USN/ ships/AO/AO-34_Chicopee.html.
Bieberstein, Art. 2004. Personal communication, March 22.
Ellis, Richard. 1994. *Monsters of the Sea*. New York: A.A. Knopf.
Ellis, Richard. 1995. Personal communication, May 22.
Holden, Constance (editor). 1995. "One Sea Monster Down," *Science*, April 14, p.207.
Mackal, Roy. 1986. "Biochemical Analysis of Preserved *Octopus Giganteus* Tissue," *Cryptozoology,* p.55.
Mackal. 1980. *Searching for Hidden Animals*. Garden City, NY: Doubleday.
Pierce, Sidney, with Gerald N. Smith, Jr., Timothy Maugel, and Eugenie Clark. 1995. "On the Giant Octopus (*Octopus giganteus*) and the Bermuda Blob: Homage to A. E. Verrill," *Biological Bulletin,* April (188), p.219.
Roach, John. 2003. "Whale-Size Mystery Creature Washes Ashore in Chile," *National Geographic News*, July 3. Posted at http://news.nationalgeographic.com/news/2003/07/0702_030702_sea devils.html#main.
Wood, Gerald L. 1977. *Animal Facts and Feats*. New York: Sterling Publishing Co.
Verrill, A.E. 1897. *American Journal of Science,* p.355.

Mysteries From Vu Quang

Linden, Eugene. 1994. "Ancient Creatures in a Lost World," *TIME*, June 20, p.52.

Rabinowitz, Alan. 1994. Personal communication, August 29.

Shuker, Karl, 1999. Personal communication, July 19.

Shuker, Karl. 1995. "Vietnam—why scientists are stunned," *Wild About Animals*, March, p.32.

Torode, Greg. 1995. "Unique Species Eaten Before Proof," *South China Morning Post*, January 7.

Fishing the Oceans

Allen, Thomas B. 1999. *The Shark Almanac*. New York: Lyons Press.

Editors of Reader's Digest. 1986. *Sharks: Silent Hunters of the Deep*. New York: Reader's Digest Books.

Kaharl, Victoria A. 1990. *Water Baby: the Story of Alvin*. New York: Oxford University Press.

McCormick, Harold W., *et al.* 1978. *Shadows in the Sea: the Sharks, Skates and Rays*. New York: Stein and Day.

Wood, Gerald L. *Animal Facts and Feats*. Sterling Publishing Co., New York, 1977.

Surviving Sloth?

Goering, Laurie. 1995. "Expert says tree sloths have a monster cousin," *Columbus (Ohio) Dispatch*, January 15.

Holloway, Marguerite. 1999. "Beasts in the Mist," *Discover*, September, p.58.

Holloway, Marguerite. 1993. "Living Legend," *Scientific American*, December, p.40.

Oren, David. 1993. "Did ground sloths survive to Recent times in the Amazon region?" *Goeldiana Zoologia* (19), (August 20).

Pearson, Stephanie. 1995. "Load the Stun Gun, Pass the Old Spice," *Outside,* November, p.34.

Volcato, Marcelo. 2003. Personal communication, September 1.

A Weird Fish

Moreno, Richard. 1995. Personal communication, February 15.

Stienstra, Tom. 1993. "Dog-faced fish leaves 'em guessing," *San Francisco Examiner*, October 3.

Fossils From Ancient Forests

Associated Press. 1994. "Prehistoric pines rock botany world," Colorado Springs *Gazette-Telegraph*, December 15, p.A16.

Boling, Rick. 1995. "Jurassic Forest," *American Forests*, March-April, p.25.

Holden, Constance. 1995. "Ancient Trees Down Under," *Science,* January 20.
Shuker, Karl. 1999. *Mysteries of Planet Earth.* Carlton Books.

Primates in the Shadows

Anonymous 2001. "Search for the Bili Ape," transcript from National Public Radio program, March 26-29.
Anonymous. 1997. "DNA Tests Identify Possible New Chimp Subspecies," http://www.enn.com/enn-newsarchive/1997/07/072497/07249713.asp, July 24.
Coleman, Loren. 1999. *Cryptozoology A to Z.* New York: Fireside.
Cousins, Don. 1990. *The Magnificent Gorilla.* The Book Guild.
Falk, Dean. 2000. *Primate Diversity.* New York: W. W. Norton.
Henderson, Mark. 2001. "Team 'find traces of Sumatran Yeti,'" London *Times*, October 27.
Martyr, Debbie. 2003. Personal communications, December 14, 15, 16.
McNeeley, Jeffrey, and Paul Wachtel. 1988. *Soul of the Tiger.* New York: Doubleday.
Shea, Brian T. 1984. "Between the Gorilla and the Chimpanzee," *Journal of Ethnobiology*, May, p.1.
Shuker, Karl P.N. 1996. *The Unexplained.* North Dighton, MA: J. G. Press.
Shuker, Karl P. N. 1996. "Going Ape," *Wild About Animals*, December, p.10.
Struthers, Elaine. 1996(?). "Koolookamba," http://www.primate.wisc./edu/pin.koola.html.
Tangley, Laura. 1997. "New Mammals in Town," *U.S. News and World Report,* June 9, p.59.
Tuttle, Russell H. 1986. *Apes of the World.* Park Ridge, New Jersey: Noyes Publications.
University of Wisconsin Regional Primate Research Center, Primate data sheets available at http://www.primate.wisc.edu/pin/factsheets.
Young, Emma. 2004. "The beast with no name," *New Scientist*, October 9, 33.

The Mysteries of Whales

Brignole, Ed, and Julie McDowell. 2001. "Amino Acid Racemization," *Today's Chemist at Work* 10(2), February, p.50.
Carl, Clifford. 1959. "Albinistic Killer Whales in British Columbia," *Report of the Provincial Museum*, p.B 29.
Carwardine, Mark. 1995. *Whales, Dolphins, and Porpoises.* London: Dorling Kindersley.
Clarke, Robert, with Anelio Aguayo and Sergio Basulto del Campo. 1978. "Whale observation and whale marking off the coast of Chile in 1964," *Scientific Report of the Whales Research Institute* (30), (Tokyo), p.117.

Dalebout, Merel. 2003. Personal communication, October 28.
Ellis, Richard. 2003. Personal communication, November 24. Also 2000, March 10.
Ellis, Richard. 1988. *The Book of Whales*. New York: Alfred A. Knopf.
Fertl, Dagmar. 1997. Personal communication.
Forney, Karen. 2003. Personal communications, August 23, October 22. Also communications from 200 and 1997, several dates.
Hain, James H. W., and Stephen Leatherwood. 1982. "Two Sightings of White Pilot Whales, *Globicephala melaenea*, and Summarized Records of Anomalously White Cetaceans," *Journal of Mammology* 63(2), p.338.
Haley, Delphine. 1973. "Albino Killer Whale," *Sea Frontiers*, March-April, p.66.
Holdsworth, E.W.H. 1872. "Note on a Cetacean observed on the West Coast of Ceylon," *Proceedings of the Zoological Society of London*, April 15, p.583.
Murdoch, W.G.B. 1894. *From Edinburqh to the Antarctic: An Artist's Notes and Sketches during the Dundee Antarctic Expedition of 1892-93*. London: Longmans, Green and Co.
Naish, Darren. 2000-2003. Personal communication, numerous dates.
Naish, Darren. 1999. "Orca-patterned monodonts?" *Exotic Zoology,* 6:1, January-March, p.1.
Naish, Darren. 1997. "Are there Narwhals in the Southern Hemisphere?" *Exotic Zoology* (4:2), March-April, p.3.
Pitman, Robert. 2003. Personal communication, October 27. Also 1997, April 3.
Pitman, Robert, *et al.* 1999. "Sightings and Possible Identity of a Bottlenose Whale in the Tropical Indo-Pacific: *Indopacetus pacificus*?" *Marine Mammal Science* 15(2), April, p.531.
Pitman, Robert. 1987. "Observations of an Unidentified Beaked Whale (*Mesoplodon* Sp.) in the Eastern Tropical Pacific," *Marine Mammal Science* 3(4), October, p.345.
Racovitza, Emile. 1903. *Expedition Antarctique Belge*. Anvers.
Cleugh, J. H., *et al.* 1994. "Cetacea: North Pacific" (description of unidentified whale), *The Marine Observer*, April, p.61.
Shuker, Karl. 1993. *The Lost Ark*. London: HarperCollins.
Shuker, Karl. 1997. *From Flying Toads to Snakes With Wings*. St. Paul, MN: Llewellyn Publications.
Sylvestre, Jean-Pierre. 1993. *Dolphins and Porpoises: A Worldwide Guide.* New York: Sterling.
Turner, Lucy. 2000. "The World's Oldest Mammal," *The Mirror*, November 18, p.14.
Urban-Ramirez, Jose. 1992. "First Record of the Pygmy Beaked Whale *Mesoplodon peruvianus* in the North Pacific," *Marine Mammal*

Science, October, p.420.
Ward, Heather K. 2000. *Cetacea*. http://www.cetacea.org.

Cryptids That Never Were

Anonymous. 1997. "One Hump or Two?" *OMNI*, March. (online edition at (http://www.omnimag.com/antimatter).
Ashley-Montague, Francis M., 1929. "The Discovery of a New Anthropoid Ape in South America?" *The Scientific Monthly*, p.275.
Coleman, Loren. 2000. Personal communication, September 6.
Cousins, Don. "Ape Mystery," *BBC Wildlife*, April 1982, p.148.
Douglas, Harry S. 1956. "The Legend of the Serpent," *New York Folklore Quarterly* (12), p.37.
Greenwell, J. Richard, 1988. "Florida 'Giant Penguin' Hoax Revealed," *ISC Newsletter* (7:4), p.1.
Greenwell, 1983. "New Guinea Expedition Observes Ri," *ISC Newsletter* (2:2), p.1.
Nickell, Joe. 1999. "The Silver Lake Serpent—Inflated Monster or Inflated Tale?" *Skeptical Inquirer*, March/April.
Wagner, Roy, Tom Williams, J. Richard Greenwell, and Gunter Sehm (various dates: correspondence) in *Cryptozoology* 2, 5, and 6.
Williams, Thomas R. 1985. "Identification of the Ri Through Further Fieldwork in New Ireland, Papua New Guinea." *Cryptozoology* (4), p.61.
Smith, Gordon, 1985. "The Case of the Reclusive Ri," *Science 85*, January/February, p.85.

Australia's Shadow Predators

Douglas, Athol. 1990. "The Thylacine: A Case for Current Existence on Mainland Australia," *Cryptozoology* (9).
Duff, Eamon. 2003. "Big cats not a tall tale," Sydney *Morning Herald*, November 2.
Healy, Tony, and Paul Cropper. 1994. *Out of the Shadows*. Ironbark Press.
Squires, Nick. "Australian 'black puma' is filmed prowling in Victoria," *Electronic Telegraph*, September 13.
Wroe, Stephen. 1999. "Killer Kangaroos and Other Murderous Marsupials," *Scientific American*, May, p.68.
Also: Various Australian news reports collected and provided by Paul Cropper.

Wolves of the Sea

Baird, Robin. 2002. *Killer Whales of the World*. Stillwater, MN: Voyageur Books.
Baird, Robin. 1997. Personal communication, March 28.

Baird, Robin, *et al.* 1992. "Possible indirect interactions between transient and resident killer whales: implications for the evolution of foraging specializations in the genus Orcinus," *Oecologia* 89, p.125.
Berzin, A.A., and V.L. Vladimirov. 1983. (translated by S. Pearson.) "A new species of killer whale (Cetacea, Delphinidae) from Antarctic waters," *Zool. Zh.*, 62(2), p.287, reprint provided by the National Marine Mammal Laboratory.
Carwardine, Mark. 2001. *Killer Whales.* New York: DK Publishing.
Carwardine, Mark. 1995. *Whales, Dolphins, and Porpoises.* London: Dorling Kindersley.
Ellis, Richard. 2000. Personal communication, March 11.
Ellis, Richard. 1988. *The Book of Whales.* New York: Alfred A. Knopf.
Ford, John, and Graeme Ellis. 1999. *Transients: Mammal-Hunting Killer Whales of British Columbia, Washington, and Southeastern Alaska.* Seattle: University of Washington Press.
Forney, Karen. 2003. Personal communication, October 22. Also communications from 2000 and 1997, numerous dates.
Hutchison, Kristan. 2002. "McMurdo whales may be unique," *The Antarctic Sun*, http://www.polar.org/antsun/oldissues2001-2002/2002_0127/whales.html.
Pitman, Robert. 2003. "Good Whale Hunting," *Natural History*, December 2003-January 2004, p.24.
Pitman, Robert, and Paul Ensor. 2003. "Three forms of killer whales (*Orcinus orca*) in Antarctic waters," *Journal of Cetacean Research and Management*, (5:2), p.131.
Rice, Dale. 1998. *Marine Mammals of the World.* Lawrence, KS: The Society for Marine Mammology.
Sylvestre, Jean-Pierre. 1993. *Dolphins and Porpoises: A Worldwide Guide.* New York: Sterling.

The Yarri

Burton, Maurice. 1952. "The Supposed 'Tiger-Cat of Queensland," *Oryx* (1), p.821.
Groves, Colin. 1992. Review of Karl Shuker's *Mystery Cats of the World*, *Cryptozoology* (11), p.119.
Shuker, Karl. 1995. *In Search of Prehistoric Survivors.* London: Brandford.
Woodford, James. 2000. "Marsupial's wimp tag now just prehistory," Sydney *Morning Herald*, February 29.
Wroe, Stephen. 1999. "Killer Kangaroos and Other Murderous Marsupials," *Scientific American*, May, p.68.

Whale of a Hybrid

Anonymous. 2000. "Loneliness might have prompted whale mis-match,"

Australian Broadcasting Company report, September 1.
Baird, Robin, *et. al.*, 1998. "An intergeneric hybrid in the family Phocoenidae," abstract, posted to MARMAM@UVM.UVIC.CA mailing list, March 12.
Baird, Robin. 1997. Personal communication, March 28.
Berube, Martine, and Alex Aguilar. 1998. "A New Hybrid Between a Blue Whale, *Balaenoptera musculus*, and a Fin Whale, *B. physalus*: Frequency and Implications of Hybridization," *Marine Mammal Science* 14(1), January, p.82.
Carwardine, Mark. 1995. *Whales, Dolphins, and Porpoises*. London: Dorling Kindersley.
Ellis, Richard. 2000. Personal communication, March 10.
Ellis, Richard. 1989. *Dolphins and Porpoises*. New York: Alfred A. Knopf.
Heide-Jorgensen, Mads, and Randall R. Reeves. 1993. "Description of an Anomalous Monodontid Skull From West Greenland: A Possible Hybrid?" *Marine Mammal Science* 9(3), July, p.258.
MICS Research. No date. "Blue Whale Research Session in Iceland with Richard Sears," http://www.roqual.com/iceland.htm.
Naish, Darren. 2001. Personal communication, September 28.
O'Neill, Michael. 1999. "DNA Breakthrough May Aid Monitoring of Commercial Whaling Ban," *BioBeat*, http://www.appliedbiosystems.com/molecularbiology/BioBeat/115-99/whale.html, January 15.
Redmond, Ian. 1993. "Beluwhales break out," *BBC Wildlife*, November, p.12.
Rice, Dale. 1998. *Marine Mammals of the World*. Lawrence, KS: The Society for Marine Mammology.
Spilliaert, R., *et al.* 1991. "Species Hybridization between a Female Blue Whale (*Balaenoptera musculus*) and a Male Fin Whale (*B. physalus*): Molecular and Morphological Documentation," *Journal of Heredity* 82(4), p.269.

The Secret of Lake Iliamna

Anonymous. 1988. "The Iliamna Lake Monster," *Alaska*, January, p.17.
Alsworth, Glen. 2000. Personal communication, November 2.
Coleman, Loren. 1999. *Cryptozoology A to Z*. New York: Fireside.
Fahey, Kim. 2003. Personal communication, May 19.
Foley, John. 1991. "Mystery monster tales keep Newhalen residents on guard," Anchorage *Times*, July 8.
Hendry, Andrew P. 1996. "At the End of the Run," *Ocean Realm*, March/April, p.52.
International Game Fish Association, "World Record Freshwater Fish," http://www.schoolofflyfishing.com/resources/worldfreshrecords.htm.

LaPorte, Tim. 2000. Personal communication, October 5.

Larson, Don. *Sturgeon Page*, htttp://www.worldstar.com/~dlarson/html/welcome.html.

Larson, Don. 2000. Personal communication, January 31.

Lew, Warner. 2000. Personal communications, September 25 and 26.

Mangiacopra, Gary. 1992. "Theoretical Population Estimates of the Large Aquatic Animals in Selected Freshwater Lakes of North America." Academic paper.

McKinney, Debra. 1989. "Believe it or Not," Anchorage *Daily News*, April 14, p.H1.

Morgan, L. 1978. "A Monster Mystery," *Alaska*, January, p.8.

Snifka, Lynne. 2004. "Monstrous Mysteries," *Alaska*, October.

Additional correspondence, comments, and research furnished by Chris Orrick.

A Kingdom of Mysteries

Anonymous. 2002. "Tuning in to a deep sea monster," *CNN.com*, June 13.

Byrne, Peter. 1990. *Tula Hatti: The Last Great Elephant*. Boston: Faber and Faber.

Croke, Vicki. 2000. "The Deadliest Carnivore," *Discover*, April, p.69.

Domico, Terry. 1988. *Bears of the World*. New York: Facts on File.

Gandar Dower, Kenneth C. 1937. *The Spotted Lion*. Boston: Little, Brown and Company.

Goodwin, George C. 1946. "Inopinatus the Unexpected," *Natural History*, November.

Goodwin, George C. 1946. "The End of the Great Northern Sea Cow," *Natural History*, February.

Greenwell, J. Richard. 1992. Personal communication, July 7.

Halfpenny, James. 1995. Personal communication, April 30.

Hawkes, Nigel. 1996. "Explorer finds giant elephants in Nepal," London *Times*, May 15.

Hichens, William. 1937. "African Mystery Beasts," *Discovery*, December.

Heuvelmans, Bernard. 1986. "Annotated Checklist of Apparently Unknown Animals with which Cryptozoology is Concerned," *Cryptozoology* (5), p.1.

Heuvelmans, Bernard. 1959. *On the Track of Unknown Animals*. New York: Hill and Wang.

Hills, Daphne M. (Mammal curator, British Museum.) 1994. Personal communication, July 22.

Keeling, Clinton. 1995. Letter in *Animals & Men*, Issue 7, p.39.

Ley, Willy. 1959. *Willy Ley's Exotic Zoology*. New York: Viking Press.

Mackal, Roy. 1987. *A Living Dinosaur? In Search of Mokele-Mbembe*. New York: E. J. Brill.

Mackal, Roy. 1980. *Searching for Hidden Animals*. New York: Doubleday and Co.

Merriam, C. Hart. 1918. "*Vetularctos*, A New Genus Related to *Ursus*," *North American Fauna* (41).

Sanderson, Ivan T. 1937. *Animal Treasure*. New York: Viking Press.

Sehm, Gunter. 1996. "On a Possible Unknown Species of Giant Devil Ray, *Manta* sp.," *Cryptozoology* (12), p.19.

Shuker, Karl P. N. 1993-2005. Personal communications (numerous dates).

Stejneger, Leonhard. 1936. *Georg Wilhelm Steller*. Cambridge, MA: Harvard University Press.

Wood, Gerald L. 1977. *Animal Facts and Feats*. New York: Sterling Publishing Co.

Note on Taxonomy

Allaby, Michael (editor). 1992. *The Concise Oxford Dictionary of Zoology*. New York: Oxford University Press.

Ferguson, J. Willem. 2002. "On the use of genetic diversity for identifying species," *Biological Journal of the Linnaean Society*, 75, p.509.

Hanken, James. 1999. "4,780 and Counting," *Natural History*, July-August, p.82.

Heins, David, and Thomas Sherry. 1997. "Species Concepts." Notes for *Ecology and Evolutionary Biology 208, Processes of Evolution*, Tulane University, http://www.tulane.edu/~eeob/Courses/208index.html.

Hutchinson, George. 1959. "Homage to Santa Rosalia, or Why Are There so Many Kinds of Animals?" Available at http://oz.plymouth.edu/~lts/ecology/santarosalia.html.

Margulis, Lynn, *et al.* 1994. *The Illustrated Five Kingdoms*. New York: HarperCollins.

Mayr, Ernst. 1996. "What is a Species, and What is Not?" *Philosophy of Science* (63), p.262. Available at http://www.aas.org/spp/dspp/dsbr/EVOLUT/mayr.ht m.

McCollough, Cherie. 2003. Personal communications, several dates.

Nowak, Ronald M. 1999. *Walker's Primates of the World*. Baltimore, MD: Johns Hopkins University Press.

Peplow, Mark. 2004. "Giant virus qualifies as 'living organism'," news@nature.com, October 14.

INDEX

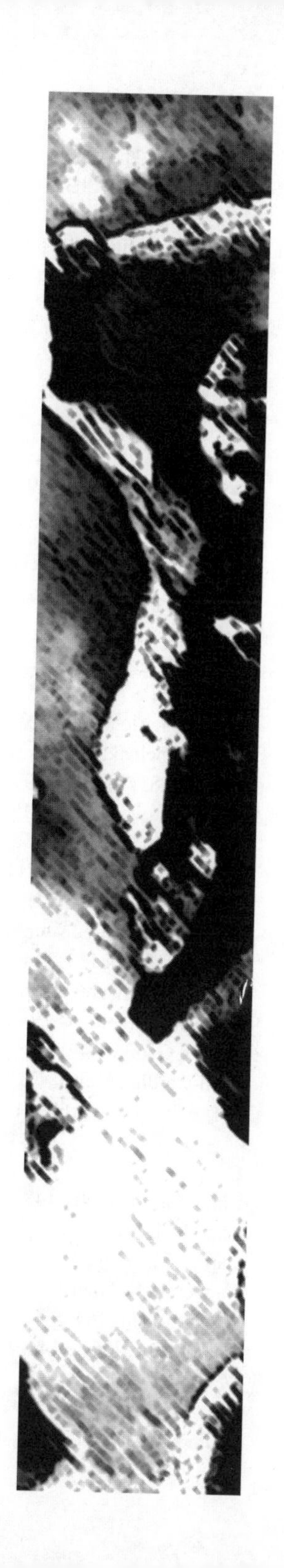

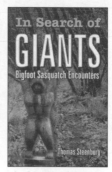

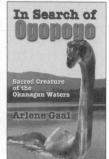